PRAISE FOR *MUSHROOMING*

"*Mushrooming* by Diane Borsato is a merry, idiosyncratic guide, charmingly illustrated by Kelsey Oseid. A sort of collector's journal of mushrooms and mushroom-related experiences, the book has entries on more than 60 North American species and is punctuated with personal essays, including one about Ms. Borsato's foray into the medicinal-mushroom world of New York's Chinatown."

—**Eugenia Bone,** *The Wall Street Journal*

"A delightful, eclectic, and enjoyable tour of mushrooms and their place in our world . . . This is a book that isn't afraid to mix the personal with the practical and will thrill many readers because of that."

—**Tristan Gooley,** *New York Times*–bestselling author of *The Lost Art of Reading Nature's Signs*

"This curious compendium of all things fungi will be welcomed by amateur mycologists, avant-garde artists, intrepid woods trampers, cautious cooks, and even armchair readers. . . . The illustrations alone, saturated with eye-pleasing, earthy colors, are worth the price."

—***Booklist***

"Absolutely terrific. This is a volume that, at first glance, is a field guide. But peering closer, as every mycologist knows, brings reward. Borsato's species descriptions are bright and lyrical; accompanying caps and stems rendered lovingly by Oseid. From art to culture to food, *Mushrooming* is a celebration of the human relationship with fungi."

—**Jonathan C. Slaght,** author of *Owls of the Eastern Ice*

"Every turn of the page in *Mushrooming* reveals surprising delights, between the eye-catching illustrations, cleverly quippy commentary, and insightful background information. Whether you're the sort to dig in the dirt in search of mushrooms, are considering your next gastronomic adventure, or are simply eager to learn more about the natural history of our fungal friends—this book will speak to the mycologist in everyone."

—**Emily Graslie,** creator of The Brain Scoop and ART LAB on YouTube and host/writer of *Prehistoric Road Trip* on PBS

"Amid 'shroom boom,' this book is ripe for the picking. . . . *Mushrooming*, which celebrates more than 120 different kinds of fungi through stories and art, takes a different tack than a traditional field guide. For one, Borsato isn't a mycologist—she's an artist. So the book explores not just individual mushrooms, but contemporary art and projects that embrace them."　　**—Globe and Mail**

"There is a worldwide resurgence of interest in fungi, so *Mushrooming* arrives at a perfect time. . . . Diane Borsato's tales of mushroom hunters, the gouache renditions by Kelly Oseid, and, yes, the culinary side of mushrooms (choose them carefully!) combine to make this book a delightful adventure from beginning to end."

—Jay Ingram, author of *The Science of Everyday Life*

"I want to live in Diane Borsato's brain. What a joy it is to experience the fruiting bodies of her deep wide intellect in this book about mushrooms, art, forests, feminism, poison, pasta, and so much more. This is my favorite kind of fence jumping, perspective shifting, deliciously illustrated, eloquently enquiring, radically alive book."

—Kyo Maclear, author of *Birds Art Life*

"*Mushrooming* is as much about identification as it is about finding wonder in all that surrounds us. Borsato frames the silent hunt through art, ecology, spirituality, ethics, and access. A guide unlike any other, this book is a must for artists, nature lovers, and anyone looking to find more magic in the world around them."

—Jen Delos Reyes, director and founder of Open Engagement

"Fungi are an order so mysterious they push our wonder and horror buttons simultaneously. In *Mushrooming*, Diane Borsato uses her radical curiosity to consider what they have to teach us about interconnection, the diversity of life, and the wildly creative possibilities of risk and attention."

—Shary Boyle, artist

MUSHROOMING

An Illustrated Guide to the Fantastic, Delicious, Deadly, and Strange World of Fungi

DIANE BORSATO

Illustrations by Kelsey Oseid

THE EXPERIMENT

NEW YORK

The Experiment, LLC
220 East 23rd Street, Suite 600
New York, NY 10010-4658
theexperimentpublishing.com

The Experiment's books are available at special discounts when purchased in bulk for premiums and sales promotions as well as for fundraising or educational use. For details, contact us at info@theexperimentpublishing.com.

Library of Congress Cataloging-in-Publication Data

Names: Borsato, Diane, author. | Oseid, Kelsey, illustrator.
Title: Mushrooming : an illustrated guide to the fantastic, delicious,
  deadly, and strange world of fungi / Diane Borsato; illustrations by
  Kelsey Oseid.
Other titles: Illustrated guide to the fantastic, delicious, deadly, and
  strange world of fungi
Description: New York : The Experiment, [2023] | Includes index.
Identifiers: LCCN 2022041056 (print) | LCCN 2022041057 (ebook) | ISBN
  9781615199587 | ISBN 9781615199594 (ebook)
Subjects: LCSH: Mushrooms--North America--Identification. | Edible
  mushrooms--North America--Pictorial works. | Edible mushrooms--North
  America--Identification. | Field guides.
Classification: LCC QK617 .B756 2023 (print) | LCC QK617 (ebook) | DDC
  579.6--dc23/eng/20220831
LC record available at https://lccn.loc.gov/2022041056
LC ebook record available at https://lccn.loc.gov/2022041057

ISBN 978-1-61519-958-7
Ebook ISBN 978-1-61519-959-4

Cover and text design by Beth Bugler

Manufactured in India

First printing March 2023
10 9 8 7 6 5 4 3

Publisher's Note

Mushroom identification is extremely complex, even for people with many years of experience. Many common edible mushrooms have toxic and sometimes deadly look-alikes. The possession of some mushrooms may also be illegal in your area.

The author, illustrator, publisher, and agents who have produced this book advise that you do not rely on this guide to identify mushrooms for consumption and do not violate laws in your area. Consume and/or possess wild mushrooms at your own risk.

Contents

How to Use This Book

THIS BOOK OFFERS INSIGHTS, anecdotes, and details about more than 120 common and charismatic fungi species in the forest, field, market—and even in the basement.

From thousands of species that are well represented across the northern United States and Canada (and in northern Europe and elsewhere globally around the northern hemisphere), I've selected an idiosyncratic bunch of common species; beloved edibles, hallucinogens, medicines, and aphrodisiacs; some specimens mythologized for their provocative shapes and colors; the enchantingly named; and many other weird and colorful personal favorites. Among the mushrooms, I've also included a few related things that might present themselves when mushrooming, including lichen, slime molds, bones, and plants. Finally, this uncommon field guide expresses my perspective as an artist and arts educator, as I have included a selection of recent, socially engaged contemporary artworks that feature fungi as a material or subject. These works variously explore themes of decay, environmental remediation, spiritual experience and/or the natural connectedness among living beings—contributing to some of the most important conversations we are having today about who we are and how we want to live in relationship with the natural world.

While I hope that beginner mushroomers—as well as advanced mycophiles—might learn new and surprising things about familiar species and find inspiration to connect more intimately with nature and one another, I would suggest you do not rely on this book (written by an artist, not a mycologist) to make accurate identifications of your specimens. There are thousands of known mushroom species and untold thousands of unknown mushroom species, so this book represents only the narrowest sampling of their diversity and splendor. Note, too, that common names are very fluid—they can be completely different in different places, time periods, and even among individual mushroomers. There can be many common names for the same fungus, and common names can broadly include several similar species. The common names I used in this book are specific to my own experience, references, and region. There is a lot of leeway and creativity around common naming, so sometimes I even dare to make them up! I strongly recommend that you refer to several other up-to-date technical field guides for fungi in your region to deepen your knowledge of mushroom identification and edibility. I have suggested other references at the end of this guide.

I also hope this book will make the case that even if you don't find an edible mushroom, or aren't sure if what you have found is edible, that the close looking, research, and even the mystery are part of the fun of mushrooming. Instead of eating a mushroom, you may just want to enjoy it for its poetic name, beauty, mythology, or specific weirdness—which is a very safe and surprisingly fruitful activity.

It's easy to love mushrooms just for being themselves—for emerging and swelling and dissolving—as an immediate expression of the weather, the season, a place and a time. It is a slow and contemplative thing to go mushrooming, letting you feel at once connected to everything and apart from the world.

Is It Edible?

THERE IS NO EASY way to know if a mushroom is safe to eat; its specific identity must be determined. Every myth, trick, or rule of thumb you have ever heard is incorrect when it comes to knowing whether or not a species is poisonous.

It takes years of foraging and mentorship to be a confident mushroomer. It requires the close study of multiple field guides and websites, along with hands-on and eating experience, to identify an edible mushroom with certainty. And even then, the mushrooms we are "certain" about may have their names changed, move into new genetic categories, or be found to cause new uncomfortable symptoms somewhere. Up-to-date advice and experience is essential to know if a mushroom is safe for eating.

A mushroom's identity must be confirmed by a knowledgeable, experienced forager that you can trust. And even if a species is confirmed to be edible, some mushrooms are toxic unless they are cooked (like morels), or toxic in combination with alcohol (like tippler's bane) or toxic until they are parboiled in multiple changes of water (like some honey mushrooms). And since mushrooms can bioaccumulate toxins from their environment vastly more than plants do, an edible mushroom growing on a roadside or in a park can be contaminated with lead, pesticides, and other pollutants and become unsafe for eating. Any mushroom can cause adverse effects and allergic reactions for some, so you must use caution by trying small quantities of a new edible specimen for the first time.

False morel,
Gyromitra esculenta

A field guide (including this one) or mobile ID app, with all the vagaries of photography, errors by the author, and limited numbers of examples to compare your mushroom with, can be very dangerous in the hands of an inexperienced forager. Mushroom types are so numerous, and can be so similar to one another and so variable across examples of the same species, that they are much harder to identify with certainty than plants. A field guide or an app are definitely **NOT** reliable tools to identify fungi on their own.

There are many more good reasons to collect mushrooms than "for the pot." Even though many of us begin looking for fungi with that in mind, I hope this book might prompt you to explore the cultural experience of the practice, enjoy time walking in the woods, and deepen your experience and knowledge.

Many of the most dedicated mushroomers eschew eating fungi at all and are strictly interested in field identification, spore analysis, and other trivia. In fact, some believe that eating your subject of interest is an affront to both the mushroom in question and the pursuit of knowledge. A foray leader once told me, when I was pestering him and eager to know what was edible, "I like birding, too, and I don't eat *them*, either."

The Art of Mushrooming

STARTED MUSHROOMING because I was offered a wild chanterelle to eat and was afraid of being poisoned. Like many beginner mushroomers, I was first interested in edible fungi—hunting with my stomach, as they say—and proceeded to inform myself about the deadly ones.

It didn't take long, though, for me to be curious about the wider world of surprising forms, textures, colors, and smells—surpassing anything I had ever encountered among works of sculpture. Fungi seemed to be everywhere once I started to look for them, vivid and amorphous, protruding from the leaf litter. I saw rings, spindles, jellies, bells, cups, and crusts. There were spheres, slimes, puffs, horns, funnels—and even ears, wings, phalluses, and tongues! I started reading field guides to familiarize myself with the common and charismatic species and found it a comfort to be steeped in these orderly texts, with their keys and categories, and the illusion that everything I needed to make nature comprehensible was inside them.

Yellow-foot chanterelle, *Craterellus tubaeformis*

I realized I'd never learn everything I needed to know from books, no matter how many I read. The language of mycology was new and strange, and the pictures were small and inconsistent. I needed to have the slimy and the smelly mushroom specimens in my hands and talk to people who were as familiar with them as with old friends. I met up with other amateurs to learn to see what they could see, and to glean all the practical ID miscellany that wasn't available in the field guides. I met enthusiasts of all ages, and from all walks of life, and have not stopped looking for mushrooms since.

The "quiet hunt," as mushrooming is sometimes known, is a joy. It is peaceful and still, and a source of the most meaningful intimacy with nature I have ever experienced. I started mushrooming as a break from teaching and the studio, but like many artists, I couldn't help but bring my whole mind and imagination to the activity. Like others, I've found this practice of looking for and learning about fungi to be a significant inspiration for creative cultural work.

I've been fortunate to be able to work as an artist, following my curiosity and immersing myself in new and unusual experiences for the past few decades. Beginning with an education in theater, and then sculpture, I've gone on to work with other artists and dancers and educators, as well as with non–arts professionals, including farmers, beekeepers, orchardists, and falconers, in dozens of projects that propose and reflect on ways of knowing and relating to things in the natural world. Unlike an art practice that results in a discrete painting or an object on a plinth, my work and the works of others, sometimes called live, relational, or socially engaged art, explores forms of experience that involve our bodies, all our senses, living materials that change in time, and the dynamics of our engagement with other people.

Bleeding tooth,
Hydnellum peckii

Students of this kind of art making from my generation (particularly in the US and Canada) looked to the works of conceptual artists from the 1970s onward—like Bruce Nauman, who crafted eccentric experiments for his own body, or Yoko Ono, who commanded the public to FLY on public billboards. We thought a lot about the raw, unfiltered gestures performed by artists like Marina Abramović, who rejected the symbolic "ketchup" used in theater in favor of drawing her own real blood. And there was Linda Montano and Tehching Hsieh in New York, who, in a most extreme example of artists making their entire lives into art, tied themselves together with a six-foot rope and lived day and night like that without a break for a full year. This generation of artists prompted those of us working now to see the world outside of art galleries and painting studios as a context for art making, and to see almost anything around us in our everyday lives as a valid material.

Together with my own students, I look at the contemporary artists influenced by these precedents: artists who walk, like Francis Alÿs tripping on drugs around Mexico City, and artists who talk, like Andrea Fraser impersonating a gallery guide to critically deconstruct the official narratives of the institution that hosts her work. We consider artists who gather together with others, like Jeremy Deller, who enlisted a conventional brass band to play electronic jazz, and Gustavo Artigas, who invited two soccer teams from Tijuana and two basketball teams from San Diego to play simultaneously in a chaotic (but not impossible!) game on the same court. And we talk about artists who reject ideas of individual authorship and mastery by collaborating with unlikely partners, as Lenka Clayton and Maggie Groat do in works of video and sculpture created with their children.

In a course I teach on contemporary environmental art in particular, artists use living materials in works they make outdoors instead of in studios and art galleries. These works require us to see our everyday environments differently and propose more reflective and new relationships to land, water, air, and the other beings we share them with. We follow the research of Tim Knowles, who makes wind drawings with trees and creates walks that follow the natural flow of water down hillsides. We admire Katie Paterson, who maps dead stars and records the sounds of melting glaciers in Iceland. We visit the work of Mike MacDonald, the Mi'kmaq artist who planted butterfly gardens full of medicines as public art, and consider Wolfgang Laib, who collects the pollen from thousands of flowers by hand to create installations of pure yellow light. There are even works by artists in collaboration with animals, like Aganetha Dyck's sculptures that are enveloped by wax comb completed by honeybees and Nina Katchadourian's playful attempts to enlist a live caterpillar as a mustache for what she calls "nature drag." Every imaginable pursuit is on the table as method or subject. There are contemporary artists who farm and who identify birds. Artists who paint flood lines and who steal ripened urban fruit. Artists who act like coyotes and who exchange soil for soup.

My students and I read the writings of environmental activists, geographers, poets and Indigenous scientists like Robin Wall Kimmerer of the Potawatomi Nation, who implores us to harvest honorably, express gratitude and recognize that all our flourishing is mutual. The expectations we have about everything in our environment are upended by these creators who propose new ways to see and know, and to be together with other beings outdoors.

But how do you practice making work like this? What do you do to develop original insights and new stories about nature? What kind of exercises support not just outdoor survival and comfort, but the kind of close looking and radical imagining that thinking like an environmental artist demands? And what can all of us learn from artists who model new modes of sensing, thinking, and being in the world? What happens if we all practice?

✦ ✦ ✦

In one of his boldest and most surprising moves among so many, the twentieth-century experimental composer and Buddhist John Cage taught both a course in music composition at the New School in the 1950s and, by his insistence, a field course in mushroom identification. Cage began mushrooming as a way to feed himself when he was starving in California in the 1930s. Learning to distinguish between edible and poisonous fungi became a fervent passion for him, and he went on to become a lay expert, a forager for chefs, and one of the first presidents of the New York Mycological Society.

In his music course, he asked students to put down their traditional instruments and listen to kitchen appliances, bicycle bells, radio station static, and paper clips, and gave them exercises in composition using scores from everyday life, like train schedules. He used found sounds and chance-based methodologies for his own compositions, and made concentrated studies of noise and silence. His most notorious score, 4'33", asks a performer to sit in front of her instrument, without playing, for 4 minutes and 33 seconds. Likewise, in his mushrooming course in the woods, he asked that students listen to the world. Once, he proposed that they try to distinguish the sounds of microscopic fungal spores, in all their different forms and sizes, falling through the air and striking the ground. And he told a researcher curious about the connection between

Parrot waxy cap,
Hygrocybe psittacina

mushrooming and music, "I have spent many pleasant hours in the woods conducting performances of my silent piece."

In addition to teaching mushrooming to musicians, he published many drawings, poems, and scores that began with Latin mushroom names, collected in his diaries and in the beautiful book *John Cage: A Mycological Foray,* which Cage made together with a botanist and an illustrator in the 1970s. But were mushrooms significantly influential to Cage's music, or to the music of his students? While he considered mushrooms to be no more or less sacred or musical than any other object, and (despite the fashion of the era) wasn't at all interested in "tripping" on mushrooms, Cage declared his love of them often. He even noted that the word "mushroom" tends to immediately precede the word "music" in dictionaries.

✦ ✦ ✦

Once you are, perhaps, dusted by the spores, mushrooms have a way of making themselves surprisingly and urgently relevant across many disciplines and spheres of life, and I can sympathize with Cage's obsessive enthusiasm for fungi. Not only do they touch many aspects of my own thinking and living, but I have also found it useful to share mushrooms and mushrooming in many forms with other visual artists and with my own students.

Whether it's part of an advanced course on food in art, walking and art, or contemporary environmental art, I always find a way to include a mushroom foray as one of our class exercises. It's easy to talk about the class themes of food, walking, or the environment while mushrooming. It's a form of direct, embodied experience and gives us a peaceful space and time dawdling in the woods—considerably less self-conscious or competitive than a seminar room—to reflect together. And it turns out that looking for mushrooms is something we can enact ritually and repeat—a practice— to support some of the most important skills necessary for artists whose materials include anything, and whose studio is the whole world.

Deadly conocybe,
Conocybe filaris

+ + +

Among the first announcements at any foray is that fungi are not plants; they contain no chlorophyll and grow in the dark. Identifying them requires a patient and rigorous sensing, especially *seeing*. You have to notice mushrooms amid the leaf litter, and then attune your vision so acutely as to see the subtlest shifts in coloration, striation, and luminosity. A single page of an identification key asks you to see if your specimen is marbled, veined, hollow, or bruised. At extremes, a key will demand you recognize the difference between *pale buff* and *pale to medium creamy-buff*. Is the stalk tapered at the apex or tapered at the base? It asks you to feel if your mushroom cap is cottony, gelatinous, grooved, or warty. Is it smooth or granular, fibrous or friable? Slimy or suede-like? And does the specimen taste peppery or acrid? Is the odor like anise or *library paste*? The list of features you are prompted to see and feel and taste and smell goes on and on. To practice mushrooming is to develop an impressive sensory literacy. And for artists working in any media, this means—almost—everything.

I tell the visual artists that I work with that looking for mushrooms is also an exercise in recognizing unseen worlds, something you can only do with an openness to mystery and surprise. You practice paying attention to the small and the ephemeral. You see what others dismiss and tread upon, what is reviled and unlovely—the slimy and degraded stuff growing in compost and rot. You see beauty in unexpected places: aphrodisiacs and medicine and food so marvelous that people will fall off cliffs to collect it. You learn to identify otherworldly portals, too, in mushrooms that conjure jewels, fairies, angels, demons, and conduits to the divine. According to ethnomycologist Lawrence Millman, the Inuit call some mushrooms *anaq*, which he translates loosely as "the shit of shooting stars."

Mushrooming is an activity that draws from numerous disciplinary conversations: biology, gastronomy, health, industrial design, history, geography, poetry, and mythology among them. While practicing it, artists and all of us learn new vocabularies to talk to one another across differences. We remember that small things must never be underestimated; they always contain multitudes.

In the woods I always direct students to look at a mass of mycelium under some bark or at the base of a mushroom to discuss how this network of fibers *is* the organism, and what we call a mushroom is only a reproductive organ, like a fruit. The fibers of one fungus can stretch out for miles, carrying nutrients to and from other creatures, doing the work of decomposition, invasion, communication, and reproduction. We can see and feel how fungi embody the connections among organisms— literally tying them all together. Mushrooming means remembering that all living things are not individual or in competition, but *interdependent*—a deep insight that radically alters the modern thinking about nature that seeks to legitimize its exploitation.

When we are sensing together among mushrooms, we can recognize the vitality and preciousness of living things, in constant tension with death and decay. And we are present, too, to the traumas of colonial and environmental violence around us. They are in our species names and field guides, and they can even be felt to reverberate right through the earth. Our fraught relationships with other people, and disregard for other forms of life, can be witnessed and reflected upon while walking together in nature. And from here, artists and all of us can think of new ways of relating to the land and living beings and create new stories to tell about them.

With thousands of known macro-fungi species, and an untold number of unknown species, mushrooming asks us, too, to bring real humility into the field. Overconfidence about the identity of a fungus, when there are so many possible microscopic differences and deadly look-alikes, can actually be fatal. I tell my students to take risks in their creative work, to push themselves and to explore the limits of what is possible in any circumstance. In the woods we practice taking risks and, most importantly, not dying from them! Artists, and all of us, can be forever amateurs among mushrooms—alert to surprise, open to being wrong and to a widening scope of possible truths. In fact, it may be that this attitude is the only one through which new ideas are possible.

✦ ✦ ✦

After John Cage's precedent exploring mushrooms and their relationship to sound and listening, it turns out that more than a few contemporary poets, artists, and architects have been inspired by fungi, exploring many of the wider themes directly connected to fungi, especially around matters of the body and mind, death and decay, and our relationships with the environment.

Indigo milk cap, *Lactarius indigo*

Gediminas and Nomeda Urbonas, for example, experiment with mycelium to grow sustainable living sculpture and building materials, and Jae Rhim Lee infuses special burial suits with spores, which, as they grow, may actually consume dead bodies and remediate the surrounding environment. Collaborations with fungi are also central to the works by Lauren Fournier and her contemporaries in a project called *Fermenting Feminism*, where artists use fermentation as a mode of resistance to mainstream food culture and social norms, and as metaphor for old ideas in explosive transformation. Carsten Höller explores a history of taking hallucinogenic mushrooms and offers gallery visitors the opportunity to experiment with *Amanita muscaria*–infused reindeer urine to experience the mind-altering, potentially spiritual effects of the fungi. Katie Bethune-Leamen makes comparisons between mushrooms and artists, as if we might all be fairy people, making fantastical things in her enormous mushroom-shaped artist studio. David Fenster and Machine Project encourage gallery-goers to recognize the marvelous and wild things growing in their city. In my own works, looking for mushrooms in globalized urban markets and museums, I'm interested in the ways that looking for mushrooms can transform our quality of attention and perception, promoting the open-ended insights that emerge when we see the world like practitioners of the "quiet hunt."

✦ ✦ ✦

When I first started to get serious about identifying fungi, a senior mushroomer, impatient with my relentless, basic ID questions told me, "When you are an infant, every animal is a dog." And some days, confronted with limitless information about thousands of variable species, I fear I haven't progressed much farther than the infant stage, still. While I have learned enough to distinguish at least a few of the mushrooms, I'm up against the boundaries of my imperfect memory, and ideas in mycology that change as quickly as the weather. I'm still full of questions and enjoying the never-ending effort to know mushrooms, and the comfort that comes when you make the world—at least a little—more ordered and known. It has given me a way of seeing and a way of being.

And walking in the woods is nice—even in the rain—and that is more than enough.

Artist's Conk

Ganoderma applanatum

ID Features: An **artist's conk** is a tough, woody bracket fungus found parasitizing deciduous (particularly beech) and pine trees. The fan-shaped bracket has no discernable stem and extends out from its host in lumpy brown waves. The visible outer edge and underside of the fungus is white. As a polypore mushroom, it has pores instead of gills. The white pore surface of the artist's conk is distinctive because it turns brown when bruised or scratched. This feature enables one to write messages and make drawings on this mushroom that are very clear and stay permanently visible even after the fungus is collected and dried up on a windowsill.

Notes: Because bracket fungi like this one are often perennial, adding new layers of tissue each year, they stick around for a long time on trees and are easy to find. Many are dull-colored, hard to distinguish from one another, and inedible. Among polypores, the artist's conk is common and widespread, and inevitably one will be found on any foray. I am grateful that I can show off the fact that the white underside of this otherwise boring-looking fungus is great to draw on, especially to artists. It's immensely satisfying to make a clean brown line on them when the pore surface is immaculate and white, but remember that it's an unforgiving substrate, and any mark made cannot be erased.

In my experience, people tend to tag this mushroom with their names or initials, or sometimes they doodle trees or draw a few mushrooms on the bracket. I have yet to see anyone do anything more surprising or conceptual with an artist's conk, though I am hoping some art student that I show it to might become inspired to do so. My friend Vito Testa, a senior member of the Mycological Society of Toronto (MST), is famous for his detailed, mycologically accurate illustrations of fungi that he will make on an artist's conk while standing in the mud, using nothing but a stick. He regularly creates these enduring works to share with his foraging friends and posts his pictures of them on the MST website. I have one of his pieces dried on my windowsill now, composed with caps and spines and corals and stars, still cleanly visible after more than a decade.

One of the other very common woody polypores I like to talk about on a foray is the **hoof fungus**, or *Fomes fomentarius*. This unattractive blackish-grey growth is famous not only because it looks like a

Artist's conk,
Ganoderma applanatum

Hoof fungus,
Fomes fomentarius

horse's hoof, but also because the frozen corpse known as Ötzi, who was believed to have lived more than three thousand years ago in the area now known as the Austrian and Italian Alps, was carrying one in his satchel when he died. It has been surmised that the hoof fungus may have been used to transport burning embers for starting fires. For this reason, *Fomes fomentarius* is sometimes called the tinder conk. I also learned recently that the myco-icon Paul Stamets wears a hat made of this tough fungus's felted mycelium.

Edibility: *Ganoderma applanatum* is not considered edible. It is so tough and woody, no one would dare try to cook and eat the flesh. However, it has been used in Chinese medicine for many years, in the form of supplements, tinctures, broths, and teas. Curiously, the primatologist Dian Fossey described gorillas picking and consuming this bracket fungus, and even savoring it, in her book *Gorillas in the Mist*.

Bird's Nest Fungi

Cyathus striatus

ID Features: Bird's nest fungi are diminutive mushrooms that grow in decaying wood, mulch, dung, compost, and manure. They are very small, each well under half an inch wide, and grow in troupes. Silvery lentil-shaped capsules are gathered in the middle of the cups and look like eggs in a nest. *Cyathus striatus* is distinctive in that the cup is "striated," or lined with regular ridges all the way around. When it rains, droplets of water land inside the nests and shoot the "eggs" out several feet away to disperse spores.

Notes: Fungi are the hidden kingdom most people step on and don't see. Bird's nest fungi are uncanny mushrooms, a hidden world within the hidden world. Even though they are common, bird's nests are so hard to find that I have only had a good look at them twice. Once was in the plastic jeweler's case carried by an accountant in the Toronto Mycological Society—a gentleman mycologist who would crawl around on the forest floor and was obsessed with collecting the tiniest fungi specimens. The second time, I was on my knees myself, wrestling some thistle out of the earth on an organic farm. Right in the mulch between beds were these tiny, sharply vivid rings around sparkling eggs, looking like they were made by miniature birds in an alternate fairy universe.

Edibility: Bird's nest fungi are not considered edible. If their habit of growing out of manure and compost isn't off-putting enough, they are so small that mycologist David Arora invents new ways to say this with every species of them (yes, maddeningly, a common name can refer to a whole bunch of species that look similar but are technically distinct), including the alliterative "much too minuscule to merit being munched on."

Bird's nest fungi,
Cyathus striatus

Black Truffle

Tuber melanosporum

ID Features: A **black truffle** is a nugget of a thing, a convoluted form covered with a dark brown crust sometimes described as "warty." This species, also called a winter truffle or a Périgord truffle, is not just warty but also wrinkly, like a baby Komodo dragon. Inside the fungus there is a tight network of pale veins that reminds me of cork. If you held a fresh one in your hand, you would find it cool and hard throughout, and unmistakably pungent, with a cloud of voluptuous gases around it.

Where I live, black truffles—originally from France, Italy, or Spain—are found floating in small jars or sliced into olive oil, honey, or sheep cheese at gourmet food shops. I have read of farmers who have been successful with, or at least hopeful about, truffle cultivation projects by planting and inoculating oak, beech, and hazel woods in the western United States, on southern Vancouver Island, and in the United Kingdom. It is true that there are naturally occurring truffles of other species in North America, though the ones we have are not as fragrant or coveted, and few of us have ever seen them— especially when there isn't a culture of looking for them buried in the dirt.

Notes: In ancient times, people thought truffles were the product of thunderbolts hurled by gods at oak trees. In the Renaissance, they were coveted and served to only the richest patrons at luxurious banquets. For millennia and across the world, the truffle has been revered as food and medicine and as an aphrodisiac. The great eighteenth-century French gourmand and pleasure advocate Jean-Anthelme Brillat-Savarin called the truffle "the diamond of the kitchen" and, contributing to its reputation as a tool of the devil to cause moral corruption, he claimed that the truffle awakened erotic dreams and "made women tenderer and men more apt to love."

I am easily enchanted by the mythology of the various choice *Tuber* species, among them the black truffle, or *Tuber melanosporum*; the less fragrant summer truffle, or *Tuber aestivum*, which is sometimes found in Britain; and of course the white truffle, or *Tuber magnatum*, the feature fungus of an entire autumn festival in Alba, in the Piemonte region of Italy—a heady place that smells entirely of Nutella, porcini, and truffles. I went for a walk with a *trifolau* (truffle hunter) there, Franco, and

Black truffle,
Tuber melanosporum

his dog, a friendly mutt named Sasha that he kept in a locked cage in the trunk of his car. In a lucrative and largely unregulated market, truffle hunters can be very competitive and territorial, and the dog of a *trifolau* can be stolen or, even more disturbingly, poisoned. Franco told me that Sasha was a graduate of the local truffle dog *university*, where dogs of any breed (and the right amount of enthusiasm) can be trained to find the precious fungi that live entirely buried underground.

Black truffles sell for several hundred dollars each, and white truffles—which are garlicky-smelling—are even more expensive, with the recent average global market price at more than three thousand US dollars for a kilogram. Big celebrity specimens—the kind with their own paparazzi and that make front-page news around the world—have been known to sell for *hundreds of thousands of dollars* each! Because of this, when Sasha stops to frantically dig, Franco catches up with her to snatch the truffle away from her sticky jaws before she devours it herself. In France, truffle hunters use sows attracted to the naturally occurring androstenol in truffles, a sex hormone that is found in the saliva of male pigs. Apparently the same hormone is found in the ripe armpits of human men, which could explain why so many people—perhaps unconsciously— are also crazed with lust for truffles, too.

Edibility: *Tuber melanosporum* is considered edible and is revered especially for its unique and savory fragrance. Truffles are the most luxurious of fungi. It's hard to say objectively if the taste and smell of truffles are worth *all* of the fuss. But like almost any experience of food, or art, I have learned that what you bring to the situation, in terms of knowledge and investment, and sometimes myth and ceremony, can greatly enhance the pleasure of your experience. But reason doesn't matter when it comes to truffles. Why not approach this rare and expensive food with some fanfare, and experience something rapturously?

This past year I made it a mission to obtain a fresh black truffle, and after a campaign of hinting (and supplying the specific GPS coordinates and market contacts) to my partner, I received one as the world's greatest holiday present. I treasured the marvelous and fragrant thing and extracted every bit of joy from that small fungus that I could. Together with my family, we smelled it and observed its veiny interior. We used a vegetable peeler to slice thin slivers of it onto creamy, barely cooked scrambled eggs drenched in butter, based on the Italian mycophile Antonio Carluccio's description of a truffle hunter's breakfast. We used it on a simple pasta and in a crock of polenta and melted Taleggio. I stored the shrinking bit of truffle in a sealed container of fresh eggs, as is advised, to subtly infuse the yolks with truffle aroma right through their shells. Just opening the bin to get an egg made me swoon from the luxuriousness of it.

In my notes from that time, I found a list of smells I collected to describe that of my fresh black truffle. It reads:

- raw egg noodles
- cold hazelnuts
- semen
- pee in soil (but good)?
- wet wood in the forest

Preserved truffles smell a little less cool and woodsy than fresh ones, but are also great and are relatively affordable. I have definitely enjoyed many pastas and pizzas cooked together with them. It pays to go to a good shop and discuss the provenance of your truffle product, since falsely labeled truffles of less fragrant substitutes are known to be rampant in the commercial market. Many truffle oils have been analyzed and found to be made entirely with synthesized fragrances and without any truffle in them at all. You might never know for sure if the truffle in your truffle oil is real. While it's better for truffle hunters and for your bank account if it is, it's possible to enjoy the aroma anyway. I do recommend, however, using truffle products sparingly and only on select occasions. It's depressing to overuse and actually get sick of such a special thing. Like with everything elusive and rare, truffles are more delicious when you have yearned for them a little.

Black Trumpet

Craterellus cornucopioides

ID Features: **Black trumpets** are thin-skinned fungi without gills, each shaped like a wrinkled cone that curls outward at its mouth edge. The interior of the cone is dark brown to black, and the exterior has a pale, dry-looking bloom. These mushrooms grow on the ground in the woods and can usually be found in groups or clumps.

Notes: *Craterellus cornucopioides* and its very similar cousins (which also go by the common name black trumpet) are so hard to find, I have heard of professional foragers using headlamps in the daylight to find them hiding in the leaf litter. I have only come across them in the wild once, when I slipped in the mud on a bank in Cape Breton and landed on my face. While recovering from the shock, I noticed I was nose-deep in a swath of black trumpets, which blended unobtrusively into the dirt. I love these charcoal-colored fungi so much that I couldn't have been happier. I shook out the earwigs from the black trumpets' hollow vase-shaped chambers and filled all my available pockets with the mushrooms. Over the years I have willfully agreed to clamber around the same slippery area, hoping to fall over again. But since I haven't yet, I reliably find dried (and occasionally fresh) black trumpets at good local grocery shops.

Edibility: *Craterellus cornucopioides* is considered to be a choice edible when correctly cooked. These are among my most favorite mushrooms to eat, and they smell strongly of earthy forest in late fall. Even though they are not as well known or prized, they are more fragrant and flavorful than many of the most famous edibles, including **morels** and **chanterelles** (pages 200 and 43); I would rank them just below **king boletes** (page 106) for their aroma and richness. They are wonderful fresh or dried and can be rehydrated without any loss of texture or flavor. Their greyish-black color, while uninteresting in the field, looks vivid and dark in any dish, contrasting with pasta or any other ingredients they are cooked with. In Quebec, black trumpets are known as *trompettes de la mort* ("trumpets of death"), making it verifiably harder to convince people in French that they are perfectly safe to eat.

Black trumpet,
Craterellus cornucopioides

Bleeding Tooth

Hydnellum peckii

ID Features: Like a **hedgehog mushroom** (page 98), this astonishing creature is a member of a group commonly called tooth fungi because it has downward-pointing spines instead of gills or pores under the cap. The common name **bleeding tooth** has, therefore, nothing to do with gruesome, horror-movie dentistry. While researchers are still not sure exactly how this feature evolved or serves the mushroom, the "blood" oozing in ruby droplets from this roughly funnel-shaped fruiting body is actually a pigmented liquid that is forced through the mature fleshy beige cap when conditions are wet.

Notes: The first time I saw a bright and unmistakable bleeding tooth, quietly bleeding its fungus blood in a patch of green maritime forest, I thought I was hallucinating. I had put *Hydnellum peckii* in the same mental category as a smiling beluga whale, or an ivory-billed woodpecker, or the nodding ladies' tresses orchid; these were rare and extraordinary things I would read about but never see in the wild. Once I calmed down and confirmed this too-good-to-be-true specimen was real, I learned it grows among coniferous trees, and while uncommon in the east, its presence is widespread across North America.

There are a few bleeding mushrooms, in fact, and if you have a penchant for wounding unknown fungi, you might be able to identify them. Among them are the bleeding milk cap (*Lactarius sanguifluus*), which bleeds when you slice it; the adorable **bleeding fairy helmet** (*Mycena haematopus*, page 29), which oozes purple juice when you pinch it; and the bleeding agaricus (*Agaricus haemorrhoidarius*), whose flesh turns red when cut.

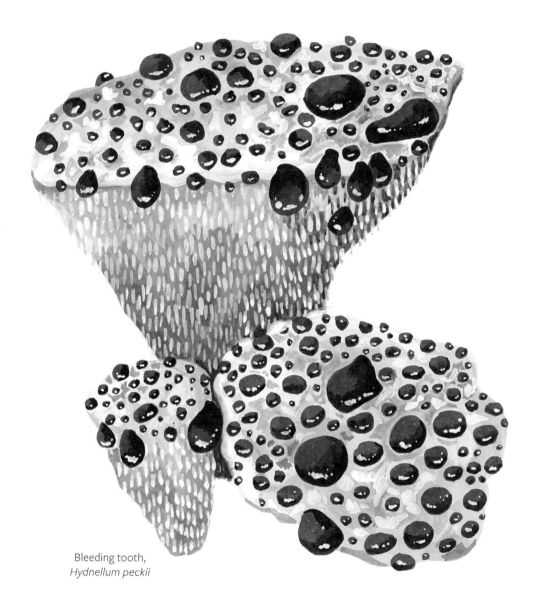

Bleeding tooth,
Hydnellum peckii

Edibility: *Hydnellum peckii* is not considered edible. On the west coast of the US and Canada, where bleeding tooth is more common, it is sometimes called the more flattering and appetizing name strawberries and cream. The flavor of this mushroom is reported, however, to be nothing like fruit or cream—rather, it is described as very bitter, and it is not recommended for eating.

One bleeding mushroom that *is* commonly consumed is the **ox tongue fungus**, or *Fistulina hepatica*, a floppy, liver-colored polypore that grows on trees. In Italy they treat it like a raw steak, and sometimes refer to the ox tongue as "poor man's meat." The chef and forager Antonio Carluccio suggests serving the ox tongue with organ meat, especially brains.

Ox tongue fungus,
Fistulina hepatica

Bleeding fairy helmet,
Mycena haematopus

Blewit

Lepista nuda

ID Features: A **blewit** is a mushroom fit for hobbits: It is plump, short, and mostly tan- or "nuda-" (the naked skin of a white European, probably) colored. It grows on decaying wood on the ground in the forest or the garden, or in the grass, sometimes in fairy rings. At various stages of its growth, the colors of its sticky cap, closely spaced gills, and stipe can change, becoming more or less noticeably purple. Sometimes the purple is subtle and other times it's as vivid as a bruise. Since a purple cast seems so unlikely in a mushroom, I have to blink away any preconceptions and focus on the fungus in order to actually see it. But the inverse is true, too, and I have to make sure I'm not hallucinating the presence of a faint lavender in a poison pie mushroom, or others, just because I would so prefer to find a blewit.

Notes: A purple mushroom is a very odd and surprising thing, like a pretty pastel Easter egg hiding in the bush. There are a few others you should know about, especially if you are looking to gather blewits for eating, including the poisonous lavender-colored *Cortinarius* species. One of these, the **purple cort** (*Cortinarius violaceus*, page 32) is also purple but has a distinctive "cortina," a curtain of hairy fibers forming a ring on the stipe, along with dark rust-colored spores.

Other purple mushrooms are more benign, closer to wildflowers than any other fungus. The **amethyst deceiver** (*Laccaria amethystina*, page 32) is a small and waxy mushroom with widely spaced gills. Its purple color is also elusive and fades with age, making it hard to identify and earning it the uncharitable name "deceiver."

And while you couldn't mistake it for a blewit, you might be happy to find a dainty **violet coral** (*Clavaria zollingeri*, page 33) for its vivid hue and oceanic intimations, or the very rare **purple prince** (*Ramaria purpurissima*, page 33), a mushroom that is larger than other purple corals, with many more fingers or branches, making it somewhat bushier upon superficial observation. The purple prince is a unique fungus that grows exclusively in old-growth forests in a few isolated areas in the Pacific Northwest. It is currently listed—as only a few fungi are—on the International Union for Conservation of Nature's Red List as a vulnerable species. Without dedicated management and conservation of the very last remaining old-growth forests, the purple prince, like

Blewit,
Lepista nuda

so many distinctive creatures, is in grave peril. If you really do come across one, you may want to photograph and record your find with a naturalist app or report it to the Fungal Diversity Survey, which can be found online. You may also want to take a knee before it instead of severing it from its mycelium and sing "Purple Rain" with real feeling in your heart.

Edibility: Blewits are considered edible when fully cooked. However, it is considered a difficult mushroom to identify, since there are other purple-, pink-, and tan-colored mushrooms that could easily be mistaken for blewits even by advanced mushroomers, including several poisonous *Cortinarius* species, *Hebeloma* species, and *Entoloma* species. Even though it is a famous edible, it takes a certain amount of familiarity and experience to forage for blewits. Fortunately, these fungi, sometimes called blue foot mushrooms, are also commercially cultivated, and can sometimes be found in gourmet food markets. They are used in many mushroom recipes—including soups, stews, sautés, and pickles—as they must be cooked to be properly edible. Note, though, that while they are sturdy and strong in flavor, they do lose their pretty hue when heated.

Purple cort,
Cortinarius violaceus

Amethyst deceiver,
Laccaria amethystina

Violet coral,
Clavaria zollingeri

Purple prince,
Ramaria purpurissima

Blue Stain Fungus

Chlorociboria aeruginascens

ID Features: This species appears in the form of very tiny teal-blue cup fungi that look more like ragged little stalked flower petals than cups. Sometimes the fruiting bodies themselves are not visible (especially in an old specimen out of season), but the fungus can be identified by a blue-green stain that looks painted in wide swaths into the grain of rotted wood. **Blue stain fungus** is common and widespread.

Notes: A chunk of odd-colored wood is easy to spot when you are looking closely for unusual things on the forest floor. Most people don't realize the distinctive color is the work of a fungus. Sometimes known as green oak, the turquoise-stained wood has been used in decorative inlaid woodwork for many centuries.

Edibility: *Chlorociboria aeruginascens* is not considered edible.

Blue stain fungus,
Chlorociboria aeruginascens

British Soldier Lichen

Cladonia cristatella

Cladonia cristatella is a good example of the kind of common, intriguing creature that may show up on the mycological foray table, even though it is technically a lichen and not a mushroom. The fruiting bodies look like tiny mint-green vases, capped with bright red flowery nubs. *Cladonia cristatella* are likely to be found mixed up with bits of moss and other lichens, making them hard to distinguish from everything else in their company. Known for its vivid red tips, this lichen is often called **British soldier**, conjuring the image of scarlet coats.

Generally lichens are found on tree bark or in moss on the ground, on old fences, and even on rocks; in fact, they are among the first family of organisms to grow right on a new lava field, creating conditions for more and more organisms, including fungi and plants, to eventually thrive. Lichens are notoriously mysterious hybrid organisms, composed of algae living in symbiotic partnership with one or more species of fungi or cyanobacteria. This unusual arrangement of species working together to make a new organism was famously described by Beatrix Potter, who not only drew storybooks for children but was also a perspicacious observer of fungi and other creatures around the Lake District in England where she lived. I always think of her as our most famous artist-mycologist; even though her exact contributions to the understanding of the nature of lichens are debated, her serious attempts to contribute to a nineteenth-century scientific establishment that disregarded women is well known.

Lichenology is still a fairly obscure and emerging discipline, with thousands of species to take into account, and where new and fundamentally world-altering discoveries are still being made. A new lichen described in 2009 was even named *Caloplaca obamae,* for Barack Obama, newly elected as US president at the time.

Lichens are famously sensitive to atmospheric pollutants and are harder and harder to find in the woods of urban centers. If you do come across some of these lichen-y British soldiers, admire them gratefully; they are beautiful, and their presence may indicate that air quality is good.

British soldier lichen,
Cladonia cristatella

Candy Cap

Lactarius rubidus

ID Features: *Lactarius rubidus* are orange-brown mushrooms that grow from rotted wood debris or on moist, mossy ground in the late fall or early winter. They are relatively small, delicate mushrooms from the *Lactarius*, or milk-mushroom, genus, and they exude a thin watery latex when the pale gills are cut with a knife. They have a faintly rough cap (sometimes described as the texture of a cat's tongue) that is depressed in the center, and a stem that is hollow when cut open.

Notes: The most distinctive thing about **candy caps** is their perfume, and identifying these mushrooms is much assisted by drying them out a little, and then entirely, to confirm you have the right species with its concentrated and enduring scent. The aroma is described to be like maple syrup, or curry—referring, I think, to the smell of fenugreek in Indian cooking, or, in my experience, the full-body sweat of a woman taking supplements of the spice to boost breastmilk supply. In fact, the association is so strong for me personally, I have had a hard time appreciating the candy cap aroma for its own sake, as it triggers unwelcome memories of my own hazy postpartum discomfort and bewilderment.

A huge range of mushroom species are described as smelling "earthy," or even more ambiguously as "mushroomy." But sometimes, in the middle of the woods in Ontario, or in any woods, the scent of a mushroom is able to conjure an entirely different world. In addition to a curry-scented candy cap, you might be surprised by mushrooms that smell like aniseed (*Clitocybe odora*), camembert (*Russula sororia*), apricots (**chanterelles**, page 43) or, curiously, even goats (*Cortinarius camphoratus*)! Depending on the age of the field guide, sometimes the smells of fungi are compared to things that were familiar to people of another time—like "fresh ground meal" (*Clitopilus prunulus*), "library paste" (toxic *Agaricus* species), "school ink" (*Agaricus xanthodermus*) or, my ambiguous favorite, "wet feathers" (*Clitocybe phaeophthalma*). Sometimes the aromas are familiar but embarrassing to declare aloud around the foray table, like "mouse droppings" (edible hen of the woods!) or "sperm" (several *Inocybe* species, among many others).

And because our sense of smell is so subjective and suggestible, people across cultures have very different comparatives for mushrooms. Smelling a mushroom to confirm its identity can be unreliable and confusing. While useful, and a delight to try, sniffing fungi should be practiced with the subjectivity of experience in mind.

Edibility: *Lactarius rubidus* is edible when correctly cooked. It is considered the one and only dessert mushroom, and while its flesh is mild or bland, a candy cap is most often used dried, steeped, and strained out to scent sweet dishes like ice cream, custard, chocolate, cookies, and other baked goods. I have even seen recipes for candy cap–infused cocktails and Popsicles.

It is hard to resist hunting for a species as famous and fancied as this sweet boundary-crossing fungus, but there are so many small brown mushrooms that candy caps could easily be confused with. These include look-alikes from within the milk-mushroom genus, and, most dangerously for beginner foragers, toxic and even deadly mushrooms from across genera, including the **deadly galerina** (*Galerina marginata*, page 125). Because of this, hunting for candy caps is only recommended for experienced, advanced-level foragers. The rest of us can try to find them dried in gourmet shops, or already baked into eccentric mushroom desserts.

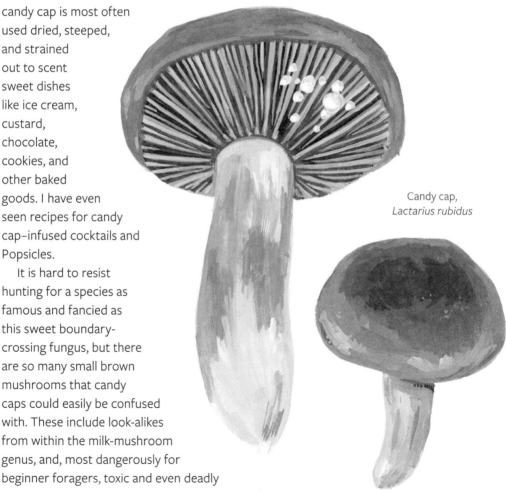

Candy cap,
Lactarius rubidus

Chaga

Inonotus obliquus

ID Features: Chaga typically grows as a bulging mass on living or dead birch trees. It looks a lot like a burnt burl, with a roughly cracked, jet-black surface that is as hard as wood. When cracked open, chaga has a golden-rust interior, streaked with paler lines and hard throughout. Do not confuse it with dead and rotting tree parts, or with **black knot** (*Apiosporina morbosa*, page 42), a fungal disease that affects many fruit trees and is also known as "shit on a stick."

Notes: Chaga is a parasite that will eventually kill the host tree, but it takes many years for this to happen. This means you can harvest chaga from the same tree over and over before it dies. However, be mindful of how much chaga you really need (not much for personal use), and when you remove it, try to do so with appropriate tools in a way that is least destructive to the tree, so you don't hasten the dying process. There has been tremendous interest in chaga and other medicinal mushrooms in recent years, and because of over-harvesting it is getting more difficult to find. There are chaga supplements, powders, tinctures, and all manner of commercial concoctions for sale, and the stated benefits of consuming it are suspiciously numerous, including treating heart disease, inflammation, aging, and cancer, and improving immunity. Like many medicinal mushrooms, it has been used in numerous cultures, particularly among Indigenous peoples, for hundreds of years. However, the health claims (as well as dosages and preparation methods) have not been studied rigorously, so the benefits or any unwelcome side effects of taking chaga medicinally may not be fully understood or known.

Edibility: When correctly identified, *Inonotus obliquus* is considered safe for consumption in low concentrations as a decoction, under the supervision of a qualified health practitioner. The body of the fungus is too hard and woody to be eaten, so when I make a chaga latte, I use a hammer to break my specimen into more practical pieces. You should only collect chaga from a living tree, and it should look fresh, clean and hard throughout. I have only tasted chaga (which is considered bitter, earthy, and even like coffee) concealed by other ingredients in medicinal broths, teas, and ciders. When I boil it, however, I notice

Chaga,
Inonotus obliquus

Black knot,
Apiosporina morbosa

it has a mild, maple syrup–like fragrance. If you are interested in the medicinal benefits of chaga (and not in harmful side effects), I would recommend talking with a qualified herbalist or other relevant medical practitioner about safe dosages and appropriate methods of preparation. It can also be safe to purchase some chaga or get advice from an experienced forager.

With this in mind, you might try my recipe for two cups of a very quick-steeped latte, made with a relatively low-concentration extraction of chaga. It takes about 25 minutes to make.

Chaga Latte

In a small saucepan, combine a walnut-sized piece (and the bits and dust) of a broken chaga with ¾ cup (180 ml) of water. Bring to a boil, then reduce heat to simmer, covered, for 10 minutes. Add 2 tablespoons maple syrup and 1½ cups (360 ml) whole milk to the pot. Bring to a boil again and immediately reduce the heat to simmer, covered, for another 10 minutes. Strain the mixture into another saucepan. Whisk the hot, sweet chaga-milk into a lofty froth. Pour into two cups and enjoy, while thinking about the many gifts of trees.

Chanterelle

Cantharellus spp.

ID Features: Most **chanterelles** are a creamy golden-orange color—like saffron stirred into custard. They have gill-like ridges that emerge from the stipe and flow seamlessly up the funnel-shaped cap. The ridges, which are not true gills, often fork as they move up the length of the fungus. Older caps are frilly around the edges, while younger chanterelles look like rounded suede nubs. These fungi can be common in some woods and may actually gleam orange-bright in dark coniferous woods—so much so that they can be seen from a car driving by. It is conventional to say chanterelles smell like apricots, but I have always wondered if the color of the mushrooms has muddled the senses of those who report this. I find the smell of chanterelles appetizing: It's buttery, eggy, and sweet, though if I have a basketful, my own senses are distracted by the excitement.

Notes: While I am skeptical about recently popular ideas of "mycelial intelligence" and the possibility that mushrooms may be using *us* in their great project of proliferation and survival instead of the other way around, chanterelles do seem to call out to you, and they are impossible to resist. They were my gateway mushroom, as they are for many people—especially in eastern Canada, where I first encountered them. More than a decade of my curiosity and learning about the kingdom of fungi— including the existence of this book—is an outcome of their allure. I'm not sure it was the result of any mushroom master plan, but I did get infatuated by these mushrooms and then fell into a steady and enduring love with them, a feeling that is with me still.

Edibility: *Cantharellus cibarius* is edible when cooked. And as they are fairly common, chanterelles are among the most beloved wild edible mushrooms collected by foragers around the world. While they are ethereal, buttery, and perfumed, and not remotely "earthy," all the classic mushroom recipes work with chanterelles. I find that a light drying for a day or two (if you can wait) is needed to reduce the amount of water in the mushrooms and make them more dense and flavorful. Then I sauté or roast them to put them on pizza or in pasta or risotto, or I bake them with cream. Freezing chanterelles can make them mushy and watery unless you precook them a bit. Drying seems to

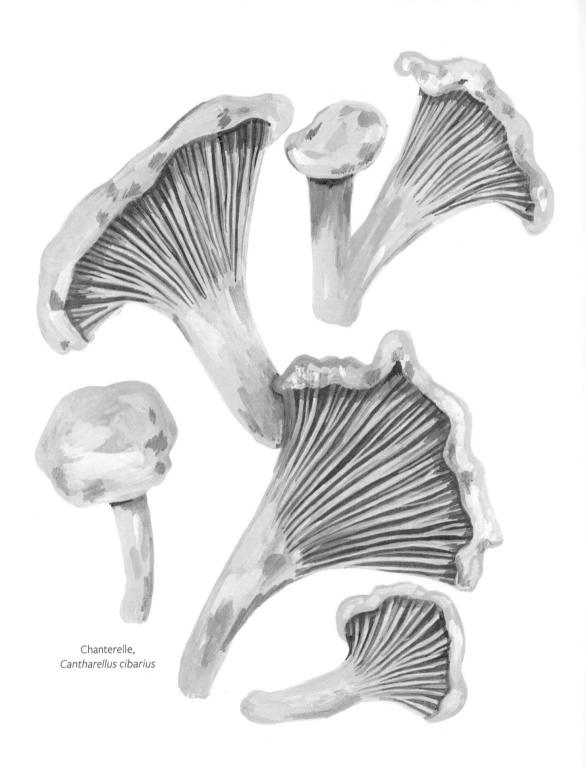

Chanterelle,
Cantharellus cibarius

make them musty, tough, and resistant to rehydration (though many people love to preserve them this way). In my experience, they are a joy to find and eat fresh for a few precious weeks in the summer. It is easy to also love the quiet work of cleaning the pine needles and slugs off them, turning and admiring and inhaling the smell of chanterelles, one by one.

As with all edible mushrooms, chanterelles are not easy to identify by any one trick; you must have knowledgeable guides and experience to identify them with certainty. There are a few closely related species to this one, including the beloved funnel-shaped *Craterellus tubaeformis,* or **yellow-foot chanterelle** (page 46), common in the west. There are also poisonous look-alikes in some regions where chanterelles are found, which can include dangerous yellow *Amanita* species. Even if you are pretty savvy, you might also confuse them with the poisonous false chanterelle, which has true gills that do not fork, and sometimes has a darker middle on the top of the cap. Or you might be tricked by the **jack-o'-lantern mushroom** (*Omphalotus olearius*, page 47), a very toxic though uncommon species that is larger than a typical chanterelle, grows in dense groups from rotting wood instead of dirt, and *bioluminesce*—meaning these eerie look-alikes literally glow in the dark.

Yellow-foot chanterelle,
Craterellus tubaeformis

Jack-o'-lantern mushroom,
Omphalotus olearius

THE CHINATOWN
FORAY

*Urban foraging project by Diane Borsato, in collaboration
with the Mycological Society of Toronto, the New York
Mycological Society, Paul Sadowski, and Gary Lincoff*

From The Chinatown Foray, Diane Borsato, 2008 (Toronto) and 2010 (New York City)

W HEN I FIRST STARTED LOOKING FOR MUSHROOMS, I started to experience an often-described perceptual phenomenon: I was seeing mushrooms everywhere. If I was outdoors, I was leaning over, inspecting mushroom-shaped stones or orange peels or (more than a few times) dried dog poop. I was seeing mushrooms in the grass or the wallpaper or the painting behind whomever I was talking to, and there were mushrooms implied by people's haircuts and hats and fabric patterns and furniture. I noticed mushrooms at the hardware store and in the supermarket. I left the woods, but when my mushrooming eyes were scanning the world, I couldn't stop seeing them, whether they were there or not, constantly, and in any environment.

On one occasion, riding my bike in the city, I had to pull over and take shelter from a sudden rainstorm in a shop in Chinatown. Passing the time among teas, medicinal herbs, and supplements, I noticed that the place was full of mushrooms . . . or was it? I could recognize dried shiitake, some jelly fungus, and obviously some brackets—but I wasn't sure about so many interesting piles and packs of dried specimens and wondered if they were plants or fungi or squid, and also if they were food or tea or medicine. Many of the species were unfamiliar to a forager in southwestern Ontario, particularly a non-Chinese one, and I realized how limited my knowledge was. I knew hundreds of mushrooms but didn't recognize fungi commonly found in a local market, blocks from my home.

Weeks later—and still thinking about the globalized, urban mycological field—I asked some friends from the Mycological Society of Toronto (many of them fluent in Mandarin and Cantonese) to help organize a mycological foray in the Chinese supermarkets and medicinal

Enoki, or velvet foot,
Flammulina velutipes

Lingzhi,
Ganoderma lucidum

shops. Foray leader Alan Gan and many others embraced the idea and proposed we conduct our intentional study closer to the suburban home of many of the Chinese Canadian members of the society: Markham, Ontario.

A weekend foray was scheduled, just as it would be for a meeting in the woods, except in this case, dozens of members of the mycological society gathered eagerly in a strip mall. With field guides and magnifying glasses, and sometimes even baskets in hand, we scanned the suburban medicinal shops and grocery stores in the same manner we would look for fungi in the forest or field.

Shiitake,
Lentinula edodes

Cauliflower mushroom,
Sparassis spathulata

I was so grateful to the many members of the society who took on new roles in this urban foray. Linda Pascali pointed out familiar specimens of dried jelly fungi and explained to non-Chinese members how to cook them. And Alan Gan translated the Chinese common names to the Latin species names for us all to identify and examine, in some cases with surprise—since our regular field guides declared that so many of the commercial species for sale in the market for consumption were "inedible"! It became more evident than ever that the distinctly Anglocentric North American perspective of many popular field guides (and, unfortunately, a few senior mycologists and foray leaders) often disregard diverse cultural uses of mushrooms and can even express hostility toward members of the immigrant communities who eat them.

Among the specimens that society members collected, purchased and laid out on the table for discussion over dim sum were **shiitake** (*Lentinula edodes*, page 51) that were sold dried in bulk and in large bags and **enoki** (page 50), which are actually a very young stage of a species native to North America, *Flammulina velutipes,* found late in the season and known as the velvet foot. There were also wood ear (*Auricularia americana*), cloud ear (*Auricularia polytricha*), king oyster (*Pleurotus eryngii*), and the fluffy, translucent fungus called the **cauliflower mushroom** (*Sparassis spathulata*, page 51)*,* which is also found wild in North American woods. There were large bags of dried and bleached netted stinkhorn (*Dictyophora duplicata*), which looked like loofah sponges. Of course, there were many examples of the bracket **lingzhi** (*Ganoderma lucidum*, page 50, also known as reishi), a shiny, banded polypore that has been used in Asia to promote health and longevity for more than two thousand years. These brackets are found in our woods, too—and even though they are used for tinctures, broths, and teas, they were among the specimens described by our local field guides as inedible.

Finally, we found examples of the infamous parasitic mushroom that is sold while still embedded in its caterpillar host, *Cordyceps sinensis*, which is used as an aphrodisiac and athletic performance enhancer (whether the evidence supports this or not) when the little infected critters are steeped or powdered and consumed.

Later, I expanded the urban forage work and repeated it in New York City, in Chinese, Japanese, Korean, Italian, and Eastern European markets across Manhattan and Brooklyn. The scope of species collected and examined convivially over dim sum in New York was extensive— including mushrooms mentioned above, and also **lion's mane** (*Hericium erinaceus*, page 119), **hedgehogs** (*Hydnum repandum*, page 98), **hen of the woods** (*Grifola frondosa*, page 162, also known as maitake), **matsutake**

Cordyceps sinensis

(*Tricholoma matsutake*, page 130), straw mushrooms (*Volvariella volvacea*), milk mushrooms (various *Lactarius* species), **blewits** (*Lepista nuda*, page 30), **chanterelles** (*Cantharellus* spp., page 43), **black and white truffles** (*Tuber melanosporum*, page 20, and *T. magnatum*), and **king bolete** (*Boletus edulis*, page 106), among so many others.

   The New York foray was conducted during a sudden winter snowstorm on streets overwhelmed by several feet of snow, with the enthusiastic collaboration of the New York Mycological Society and esteemed guest mycologists Paul Sadowski and the late Gary Lincoff.

Crown Coral

Artomyces pyxidatus

ID Features: Coral fungi are distinctive, with smooth forking branches that appear to grow in an upward direction (as opposed to tooth fungi, which feature exclusively downward-pointing spines). They look a lot like the sea creatures they are named for and come in a wide range of colors, including white, cream, yellow, orange, pink, red, and purple. The **crown coral** has beige-colored flesh and is fairly common, growing from decaying wood. The distinctive feature of these is that each upward-pointing branch ends in three or six points, like a little crown at the tip. They are also notably peppery if nibbled (and promptly spat out).

Notes: Because coral fungi are so similar to one another, and it's very difficult to distinguish the correct species without a microscopic view of their spores, Canadian and American guidebooks tend to advise caution around fungi in this wider group and do not recommend them for foraging and consumption. Those who do eat them here describe them as having an interesting texture but a bland taste and say they are hardly worth the trouble. Though, as with many fungi we are afraid to taste here, several corals are collected for the table elsewhere around the world. If you want to taste coral fungi among people who know well how to identify, collect, and cook these and other mushrooms many are afraid to eat in the US and Canada, some culinary adventure travel is in order.

Edibility: *Artomyces pyxidatus* is considered edible when fully cooked. There are numerous coral mushrooms, and it can be complicated to distinguish them from the crown coral. After consulting numerous field guides and websites, I am more confused about the identities, names, and edibility of coral fungi than I was before I looked them up. Beware: some of the other look-alike corals in various colors and shapes are poisonous and associated with stomach upset.

Crown coral,
Artomyces pyxidatus

Dead Man's Fingers

Xylaria polymorpha

ID Features: Dead man's fingers are irregularly shaped club-like mushrooms that survive on rotted wood. The rotting process initiated by this fungus is so strategic and particular that a Swiss scientist has been inoculating wood with a *Xylaria* species to lower its density and improve its acoustic properties, making it better for use in violin making. You can find these fungi in pairs or in a group of several, sometimes fused together at the base. They can be discovered all year long, in various stages of development—lighter in early spring and darkening to brownish and black by summer. The surface of the fungus is hard, matte, and faintly porous. *Xylaria polymorpha* and several of its look-alike cousins look and feel like tough charcoal.

Notes: I count on foray participants to find this mushroom, which is fairly common, and then I relish announcing its suggestive name. Occasionally you may find a cluster of five gnarled *Xylaria* that comes out of the woods like a dead man's whole hand. I think of a poem by the Swedish poet Tomas Tranströmer, "Sketch in October," in which he described seeing mushrooms in the grass, "like the help-seeking fingers / of someone who sobs down there in the darkness"; he then points to the persistent reminder of all saprophytic fungi: "We are the earth's."

Edibility: Dead man's fingers are neither appetizing nor edible.

Dead man's fingers,
Xylaria polymorpha

Deadly Webcap

Cortinarius rubellus

ID Features: Deadly webcaps are medium-size terrestrial mushrooms that grow in the woods in association with oak and other deciduous trees. They have reddish, fox-colored caps that have a nipple-like point in the center. The gills are also reddish brown, and you may notice a spray of yellow-brown spores on the stipe, or in a spore print. Their stipes are also yellowish brown to brown, and both the cap and the stem look and feel covered in tiny fibers.

Notes: *Cortinarius* mushrooms are distinguished by a cortina, which is a curtain-like ring of fibers that can be found protecting developing gills in young mushrooms or hanging around the top of the stipe when the mushrooms are more mature. When you find the fibers still attached to the cap, they look like a spiderweb-thin work of thread-and-nails art, like the optical-experimental kind that was popular in the 1970s. In the deadly webcap, for some reason, the ring of threads tends to vanish early and is not often visible anywhere on the mushroom.

Edibility: *Cortinarius rubellus* is a deadly poisonous mushroom that should not be eaten. It is said that this specimen smells like radishes. You should never nibble it, considering it contains very high levels of orellanine, a poison that causes unpleasant symptoms after a few days and ultimately kidney and liver failure. While some individuals have been saved by organ transplants, many have not. Consuming even a small quantity of this mushroom can be fatal. In this huge genus, there are many poisonous and deadly webcap species. It is widely recommended that no *Cortinarius* species be eaten.

Deadly webcap,
Cortinarius rubellus

Death Cap

Amanita phalloides

Death cap,
Amanita phalloides

ID Features: The **death cap** has many of the distinctive features of the *Amanita* genus, including a bell-shaped cap that opens out quite flat, a long slender stipe with a thin floppy ring near the top (formerly used to cover the developing gills) and a bulbous base wrapped in a papery sheath. When young, *Amanita* mushrooms are enclosed in a white sac known as the universal veil, and traces of this veil are left on the base of the mushroom and sometimes on the cap itself. While the colorings on poisonous mushrooms are not always alarming or obvious, a death cap appears as a sickly olive green.

Notes: It is so easy to mistake a big, meaty *Amanita* for an edible mushroom that most of the mushroom poisonings in the world are caused by fungi in this genus. Even for the initiated, telltale rings fall off and bulbous bases get broken and lost. In addition to other ID traps, young *Amanitas* actually look similar to edible puffballs; so any suspected puffball should be cut open to check for the silhouette of an *Amanita* in development.

Edibility: *Amanita phalloides* is a deadly poisonous mushroom that should **NEVER** be eaten in any quantity. It is crucial to identify your mushroom species and to know poisonous look-alikes very well in order to collect mushrooms for the table.

SLUDGE is the morbidly comical acronym for Salivation, Lacrimation, Urination, Defecation, Gastrointestinal distress, and Emesis—the symptoms of mushroom poisoning. The descriptions remind me of what it must be like to be electrocuted, or worse: sweating, salivating, vomiting, severe cramping, bloody diarrhea, and so on. Some people may experience a period of relief and feel better after the initial gastrointestinal symptoms, but it is only the beginning of the real problem, as the amatoxins in a death cap go on to cause major kidney and liver failure and then poison your other major organs, muscles, and brain.

While there is no antidote to poisoning by death cap, there are protocols and treatments that, if performed within a short period after ingestion, have reduced mortality to (by recent reports) around 20 percent. Otherwise it is described variously as an abject, slow, and agonizing way to die. Because of this, it is recommended that you seek emergency medical care if you have consumed any quantity of this mushroom—and not wait for symptoms.

Mushroomers sometimes tell adventurous newcomers, "There are *old* mycologists and there are *bold* mycologists, but there are no *old bold* mycologists." They also say, "All mushrooms are edible. *Once.*"

Delicious Milk Cap

Lactarius deliciosus

ID Features: The **delicious milk cap** is a very pretty mushroom with a variable color on its stipe, gills, and slimy cap, including shades of tan to greenish to carrot orange. Like other *Lactarius* species, these fungi often have funnel-shaped caps with concentric rings on them and a depression in the middle. They may also have polka-dot spots, known as potholes, on the stipe. While there are a few orange milk caps, *Lactarius deliciosus* is one of the only ones that will exude an orangey-red latex when sliced with your mushroom knife. You will find, however, that the entire mushroom will stain green wherever it is handled, bruised, or cut.

Notes: Milk mushrooms, including the delicious milk cap, can be very confusing to determine the exact species, since many are similar and require close examination of the color of the cap, the color of the staining, the color of the milk, and then how the milk changes to another color in time. *Lactarius thyinos* is an example of a very similar species with all the same visible features but without the green staining when the mushroom is handled. While *Lactarius thyinos* is also edible (and often preferred for eating), there are other delicious milk cap look-alikes that are toxic, mainly from the *Russula* genus. Depending on your foraging experience, this is a mushroom that should be collected for eating only by advanced mushroomers.

Edibility: *Lactarius deliciosus* is considered edible when cooked. Even though these fungi are very popular in European markets and are widely consumed in Spain, where they have the appetizing common name saffron milk cap, North American mycologists like to note that the delicious milk cap is not so delicious. It is described as slimy, bitter, and grainy, and many suggest that *Lactarius deliciosus* may have been misnamed by Carl Linnaeus himself. I do find these and *Lactarius thyinos* to be bland in flavor but pretty in color when cooked. I would recommend using them in small quantities for color with other, more flavorful specimens or with lots of lemon, herbs, and cream.

Delicious milk cap,
Lactarius deliciosus

Destroying Angel

Amanita virosa and *A. bisporigera*

Destroying angel,
Amanita virosa and
A. bisporigera

ID Features: The **destroying angel** is pure chalk white, often found in pristine condition, without any tiny holes bored by maggots or bite marks from rodents; this fungus can be as immaculate as a wedding dress. Characteristic of the deadly *Amanita* genus, it has a bulbous base sheathed in a cup structure called a volva, and needs to be carefully dug out of the earth so as not to damage these critical identification features. From base to top, a mature destroying angel can be 12 inches tall, clean, symmetrical, and gleaming among the leaf litter. A tissue-like ring called the annulus dangles serenely from the long thin stipe, just under the gills, which are tightly packed under the cap. A bit of wind can sometimes blow the ring away, leaving a naked stem, which can be deceptive to an inexperienced forager.

Notes: George Barron, the senior professor emeritus of mycology in Ontario, describes the destroying angel as both "the prettiest and the deadliest" of local species, the original disastrous woman of mushrooms. I always tell art students this anecdote at forays to ask them to consider "pretty and deadly" as an example of opposing forces to mobilize when you want to create tension or make something irresistibly beautiful. *Amanita virosa* is fairly common in Ontario. It was the first deadly poisonous specimen I ever lifted out of the ground. I held up the large intact mushroom to admire its perfection, and I could feel it pull the heat out of my hand. A senior mycologist told me to get it out of my basket. Can it be used as a weapon, I wondered? Who would be liable (he was blinking and nervous and averting his eyes) if it were?

Edibility: Like many *Amanita* species, *Amanita virosa* (and its nearly indistinguishable cousin, *A. bisporigera*) is deadly poisonous and should **NEVER** be eaten. A destroying angel can be mistaken for a few edible species of seasonal fungi—like the handsome horse mushroom (*Agaricus arvensis*) or the discreet naucina (*Leucoagaricus leucothites*). In fact, it can be mistaken for any white, gilled mushroom, and serious foragers should be familiar with it. Young *Amanitas*, still developing in their egg stage and covered by a universal veil, have also fooled foragers looking for small edible puffballs. I have heard people say that all puffballs are edible, but they are not. Not only are there poisonous puffballs, but a spherical white fungus may not even necessarily be a puffball. If you cut it open and see the burgeoning outline of a bulbous base, stem and cap, you don't have a puffball at all— you may have a young destroying angel

Grisette,
Amanita vaginata

going into your pot, or one of its equally malevolent *Amanita* cousins, a **fly agaric** (*Amanita muscaria*, page 84) or a **death cap** (*Amanita phalloides*, page 60).

These mushrooms account for a majority of mushroom poisonings around the world. It is reported that even half a cap can cause death in adults after extreme vomiting, convulsions, diarrhea, and more. Victims are sometimes injected with vitamins, penicillin, and high doses of milk thistle extracts through IV, which, if performed hours after ingestion, has sometimes saved lives. I have heard reports that the destroying angel was delicious, fried carelessly with butter—except that the exclusive sample of diners who have had this experience die in too great a number for these reports to be definitive.

Consider admiring species from the *Amanita* genus for their sheer exquisiteness, like one of my favorites, the ringless **grisette** (*Amanita vaginata*, page 66), a tall beauty, perfected as if by sheer pantyhose, and with delicate striations on its parasol-like cap. Or look for the pinkish blusher (*Amanita rubescens*), which is distinctive among *Amanitas* in that its flesh reddens with age and when cut. Even when some of these *Amanitas* are considered nontoxic, and a few provocative myco-adventurers do try them, I *strongly* discourage anyone from sampling any *Amanitas* in North America, even the reportedly edible ones, since they are so easy to misidentify, with potentially grave consequences.

Blusher,
Amanita rubescens

Dog Vomit Slime Mold

Fuligo septica

Dog vomit slime mold,
Fuligo septica

Dog vomit slime mold has one of the most impressively disgusting names in any of my field guides to fungi, even though it isn't a fungus at all. Slime molds, like this furry yellowish mass that will grow in garden mulch or on decaying matter, are often confused with fungi or plants, even though they are actually more comparable to animals, as creatures that need to forage for their food.

Slime molds are composed of single-cell creatures like amoebae, who gather together in a sort of greedy unified blob and travel around hunting for decaying matter, bacteria, and other microorganisms to consume, at the documented speed of approximately one millimeter per hour. Try to imagine the uncanny spectacle of a large pile of vomit ambulating like a tortoise around your garden! A stunning BBC documentary tries to make this felt by showing a time-lapse video of a slime mold moving along like a mass of pulsating veins to a moody soundtrack of sticky smacking sounds and trip-hop music. Waves of plasmodium—the body of the slime, which the narrator describes in dramatic tones as "a relentless, shape-shifting yellow goo"— coats and consumes everything in its path.

There are numerous species of slime molds, including one of my favorites, the **wolf's milk slime mold**, or *Lycogala epidendrum*; when it reaches its fruiting phase, it looks like a mass of soft, peachy-pink spheres. I love the mysterious name of this organism and wonder what the milk of wolves might actually look like. How did this mold acquire this very poetic reference? Unfortunately, there is nothing poetic about the numerous unnerving YouTube videos of men in the woods eagerly poking these spheres with sticks and mashing them gratuitously with their fingers to show the curious viewer how they are filled with a pink, milky pus. You might enjoy this experiment yourself, or you might be too late, and find wolf's milk slime mold in its more mature stage, when the spheres are a silvery brown, filled instead with a sticky, dark, spore-filled paste.

People are so fascinated by these strange creatures that I know of more than a few who keep them as pets. Richard Aaron in Toronto keeps an impressive collection of tiny samples in his freezer (semidormant, I assume), and the English mycologist Bryce Kendrick feeds his specimens porridge oats, while he studies their movement, memory, and even intelligence. Researchers have shown that slime molds like *Physarum polycephalum* (sometimes known unofficially as "the blob") can find the most economical paths to their prey, and this behavior can be manipulated to watch them make decisions and solve mazes. Famously, this species was used to draw the optimal network between rail stations in Tokyo.

Wolf's milk slime mold,
Lycogala epidendrum

Doll Eyes

Actaea pachypoda

Doll eyes,
Actaea pachypoda

This irresistible plant grows across North American woods and fruits in the late summer and early fall, just when mushroomers are scanning the ground for elusive, or remarkable, specimens. **Doll eyes** almost always ends up on the foray table, for its brilliant fuchsia stem and shiny white berries—each with a black pupil-like calyx looking back at you, unblinking.

Despite its beauty and resemblance to doll parts, be careful not to handle the specimen or share it with children! This is an extremely poisonous plant if eaten, in some cases even fatally so, and it can cause serious skin irritation when handled. It is also known as white baneberry; in botanical naming, "bane" refers to dangerous and toxic specimens. At every foray, someone asks if it's safe to touch a deadly poisonous mushroom, and while I reassure them that it is, I always add that the real danger is touching the toxic plants just like this one, as well as poison ivy, wild parsnip, giant hogweed, and many others.

Incidentally, some of the other nonanimal creatures that look back at you while mushrooming are the trees. My friend, photographer Robyn Cumming, who is preoccupied by representations of faces—like natural faces with eccentric

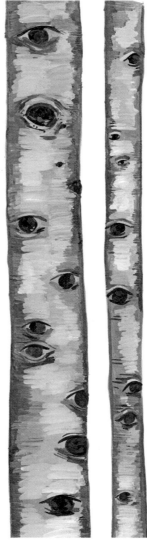

Paper birch,
Betula papyrifera

teeth and faces constructed by everyday objects—tends to have an overactive sense of pareidolia. She sees faces and their disembodied parts in random, ambiguous markings and forms. Spending any time in the woods with her, you can't stop seeing them, too—the wide, expressive eyes, especially—surveilling you from every corner of the woods, right through the bark of **birch trees**.

Domicile Cup

Peziza domiciliana

ID Features: These *Peziza* species are much like the common thin-fleshed tan or brown cups you might find in the woods. They are distinctive because of where they are found—notably on dirt floors, in garages, and in basements, sometimes growing right up through cracks in concrete. The medium to large cups are wavy, irregularly shaped, and sometimes flattened and split against the ground.

Notes: When my partner and I were looking for a house to buy in Toronto several years ago, we followed the real estate agent on tours through the various possible homes we could afford. I shrieked when we arrived in one musty basement bathroom, because in front of us were more than seven large, thriving domicile cups growing between the bathroom tiles, just under the toilet! At the time, I was developing my mushroom obsession and avidly memorizing Latin species names. I shocked the already horrified realtor by declaring the disgusting eruption *"Peziza domiciliana!"*—the very first, and since then only, example I have seen of the species.

Edibility: Domicile cups are neither appetizing nor edible. If you breathe on a cup fungus like these, though, you will not only surprise and confuse your realtor, but the mushroom may puff out a cloud of spores back at you in response, as a sac fungus like this one would properly do into the wind.

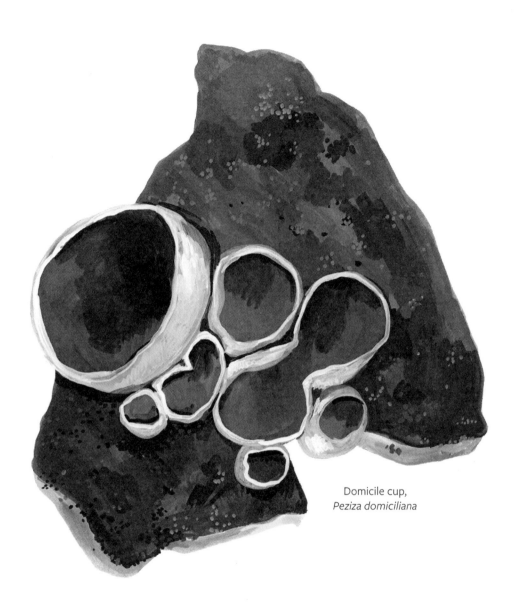

Domicile cup,
Peziza domiciliana

Earthstar

Astraeus hygrometricus

ID Features: While there are many mushrooms called **earthstars** from several genera, the *Astraeus hygrometricus* is distinguished by the many brown, pointed petals (known as rays) that radiate out from the pale greyish puffball at the center, resembling a flower. The surface of the petals is often crackled and veiny. When the spores are mature, the puffball opens a pore at its center, through which dusty brown spores puff out. Earthstars are widely distributed across North America and can be found in wet or dry fields, woods, trail sides, and wasteland areas.

Notes: There are earth balls, earth tongues, earth fans, and earth nuts among fungi growing from decomposing plant litter, though none are as novel and unexpected as the earthstar. And as if a puffball, crafted like a child's sculpture of the sun, weren't interesting enough, among the many examples of this type of fungus there are beaked earthstars, acrobatic earthstars that seem to stand like ballet dancers *en pointe* and this hygroscopic specimen. *Astraeus hygrometricus* is sometimes called the barometer fungus because it detects the levels of humidity in the air and opens its petals when conditions are wet to let raindrops pelt out its spores. In dry conditions, the petals close tightly up around the belly of this ascomycete, or stomach fungus. An earthstar is as evocative as it is strange, being at once terrestrial and celestial: a star on earth.

Some, like ethnobotanist and countercultural icon Terence McKenna, took such an idea quite literally, positing that mushrooms were sent by extraterrestrials to replicate on this planet. In the 1990s, McKenna popularized methods for home-growing hallucinogenic *Psilocybe* species and claimed that there was evidence that mushrooms like these contain a molecular fingerprint proving they were from out of this world. He proposed that mushrooms might have penetrated our environment to probe our world for life, and even that spores—which can survive in all manner of hostile conditions, including in space—may have traveled here to inhabit our brains, creating unique connective conditions that produced human intelligence.

While his theories were never embraced by mainstream researchers, I can understand the impulse to describe

Earthstar,
Astraeus hygrometricus

fungi as alien beings. They don't act like plants or animals. They grow in the dark and digest trees. Sometimes they are purple or blue or slimy, or bright red and penis-shaped. They exude juices, decompose the dead, burst out of the heads of caterpillars and sometimes make people see god and declare absolute oneness with the universe. But to me these things are less compelling as proof of aliens, and more as evidence that things on earth are wondrous and relentlessly surprising. And the more I look for things like earthstars and learn about their forms and attributes, the more convinced I am of our opportunities to behold the mysterious and the divine, right here in the earthly dirt.

Edibility: *Astraeus hygrometricus* is a tough and unappetizing fungus, and not considered edible.

Eyelash Fungus

Scutellinia scutellata

ID Features: Eyelash fungi are shaped like very small dark orange or orange-red cups, fringed in black hairs, looking distinctively like they are trimmed all around in false eyelashes. They are only about a centimeter in diameter each, though they appear together in large groups, vivid and bright against the decaying wood they fruit on.

Notes: Once you know what you are looking for, eyelash fungi are fairly common and easy to spot from afar. With their keen close-range vision, kids can see the fringe of lashes around the small cups without the magnifying loupe far-sighted adults might require, and the weird surprise of this makes these mushrooms popular with children. It is also a delight—for any of us—to announce the alliterative and delicately percussive name of these fungi: *Scutellinia scutellata*.

Edibility: It may go without saying, but a tiny, hairy eyelash fungus is not considered edible.

Eyelash fungus,
Scutellinia scutellata

Kombucha Mother Tarot Reading,
Lauren Fournier, 2014

FERMENTING FEMINISM

*Curatorial project by Lauren Fournier, in collaboration with
many other artists and fermentation practitioners*

M Y FAVORITE SCOBY (Symbiotic Culture of Bacteria and Yeast) lives behind the bar at the local Russian banya in Toronto. According to the proprietor, Valentina, this impressive, platypus-sized blob has been alive and exchanging gases for as long as her own adult children. She feeds it tea and sugar to get the popular by-product of its fermentation work: tart, delicious kombucha.

Fermentation is the work of yeasts—which are microscopic fungi—working together with bacteria to break down sugars and release nutrients from foods. It is perhaps the most familiar work of fungi, and inextricable from our everyday lives. Not only do we use yeasts intentionally to make the preserved food we love, like cheese and pickles, as well as beer, wine, cider, and other alcoholic beverages—these tiny fungi also live inside us, helping to digest everything we consume right in our own guts. In fact, yeasts are the fungi that live all over our bodies— on our skin and eyelashes and hair—along with a village of other micro- organisms (known as a microbiome) that we have evolved together with, and that we wouldn't survive without.

Lauren Fournier, artist and creator of the ongoing curatorial project called *Fermenting Feminism*, describes her own gelatinous "mother" (as SCOBYs are often called) floating in a crock of sour liquid to be "as big as a placenta." In a surprising silent video, she sits at a table lit with candles, across from a jar sitting alone on a stool. She lays out a series of tarot cards and while looking straight at the floating blob, appears to earnestly give her kombucha mother a tarot reading. Later in the piece, another performer takes her "mother" out for a walk in the park. She holds up the jar to see statues, chats with it on a park bench, makes it a daisy crown and cradles it like a baby. With ongoing references to arcane and feminist symbols, Fournier's video asks us to consider forms of intuition and knowledge practiced (often) by women and also to reflect on a weird, almost familial intimacy that we can experience with our non-human kin. After all, what are the limits of love? And how might we expand feminist ideas with this intimate connection to non-human creatures as a starting point?

As an avid fermenter herself, Fournier was curious about the ways artists, feminists, scientists, witches, brewers, and others take up fermentation as part of their material practice. In a series of ongoing exhibitions, performances, and public programs in venues from Kansas City to Vancouver to Berlin to Copenhagen, Fournier has gathered together practitioners local to each site. She wanted to see how the complex and mysterious workings of billions of interconnected microbial organisms, with fungi and bacteria among them, might be a rich trove of insight for thinking about ourselves in more expansive and inclusive ways, and about how to live together with other beings and our wider environment.

She found Benedictine nuns who craft their own unpasteurized, illegal cheeses, and queer homesteaders who make their own food and alcohol as an expression of bodily agency and in defiance of mainstream economic systems. She also included the works of artist and fermentation practitioner S. E. Nash, who makes chunky, unsettling sculptures of putty, paint, and burlap with bubbling jars of kimchi, sauerkraut, and jun tea built right into them. Audiences can eat the living, evolving foods that are made in the artworks while gathered in the art gallery. Not only does the subversive work break all the conventions of environmental purity in museums, according to Nash, but it also generates a space for thinking about the connections between symbiotic organisms, family history, and the artist's own transgender experience. Fournier presented the work of Alice Vandeleur-Boorer and Tereza Valentová, who, in defiance of powerful taboos around the bodies and bodily fluids of women, culture their own vaginal *Lactobacillus* bacteria with milk in wild ferments they call vaghurt. In Fournier's catalogue about the project, the artists describe their material as being as provocative as it is precious: "Somewhere in the middle of lust and nausea, yummy and yuck, of fresh and spoiled we found a warm and moist feminine treasure." Participants are invited to contribute their own flora, and protocols for making your own vaginal yogurt are available on their website, in an effort to welcome all kinds of radical fermentation.

In the works of *Fermenting Feminism*, contamination becomes liberation and collaboration; fermenting foods become a form of political resistance to the oppressive regulation and sanitation of our bodies, and to the globalized, capitalistic economies that determine so much of the trajectory of our lives. Fermentation also models collaboration across forms of difference that even transcend species. As Fournier describes it, "sour and effervescent, ambivalent and refreshing, multi-sensorial and potentially explosive, *Fermenting Feminism* taps into the fizzy currents in critical and creative feminist practices today." It is a way to acknowledge vibrant alternative cultures, in both the philosophical and the microbial sense of the word.

Field Mushroom

Agaricus campestris

ID Features: The **field mushroom**, sometimes called the meadow mushroom, is a squat specimen with a short, smooth stem that narrows slightly at the base. It has a white to cream cap, sometimes with brownish scales, especially in the middle as it ages. It may also show leftover tissue on the edge of the cap and some veil fragments, making a flimsy—sometimes disappearing—ring around the stem. Its gills are free, meaning they are not merged or running down the stem. The gills are a dusty-rose color when the mushroom is young, though they turn dark purple-brown, like the spores, when the mushroom is mature. Field mushrooms grow in fields and pastures, as well as in backyard grass.

Notes: Field mushrooms are familiar to many, not just because they are common and widespread, but because they belong to the large *Agaricus* genus of brown-spored mushrooms. Among these close relatives are the most recognizable supermarket mushrooms, including the little white button, the brown cremini, and the famous hamburger substitute, the **portobello** (*Agaricus bisporus*, page 82). Each of these three is simply a different developmental stage of the same species, *Agaricus bisporus*. The field mushroom shares an odor with these others, like moist earth and black licorice or aniseed.

Another famous relative of the humble field mushroom is the sidewalk mushroom (*Agaricus bitorquis*), famous for growth so forceful it can bust up through concrete, ruining more than a few newly paved driveways and tennis courts.

Edibility: *Agaricus campestris* is considered edible. While this mushroom and a few of its edible cousins are very familiar to some foragers who know where to regularly collect them, they are difficult species to identify with certainty. Several deadly mushrooms, including *Amanitas*, share a similar ring around the stem, and several brown-spored mushrooms that resemble field mushrooms, including *Cortinarius* species, have caused severe illness and death.

Many of the mushroom poisonings in North America and the UK are caused by the toxic look-alike *Agaricus xanthodermus*, known as the **yellow stainer** (page 83). It is easily confused with the field mushroom because it is also a white-fleshed mushroom with a stem ring and pink gills

Field mushroom,
Agaricus campestris

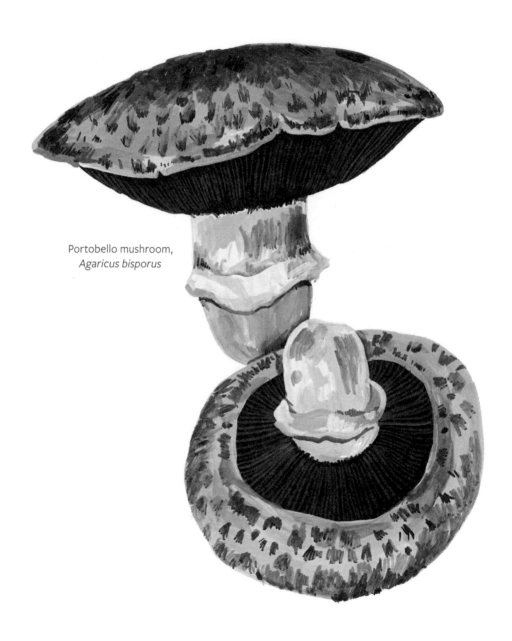

Portobello mushroom,
Agaricus bisporus

(that turn brown). The yellow stainer is known for turning yellow when bruised or cut, especially at the base of the stem. And it stinks; mushroomers have described it as smelling variously like antiseptic, iodine, or ink. There are at least three other poisonous *Agaricus* species look-alikes that are very hard to distinguish from the field mushroom that share this odor, but don't necessarily turn yellow. Sharing features with so many poisonous species, this humble-sounding mushroom is rightly placed in the advanced/experienced category of foraging.

Yellow stainer,
Agaricus xanthodermus

Fly Agaric

Amanita muscaria

ID Features: When mature, a **fly agaric** is a stout, impressive mushroom with the signature features of an *Amanita*, including a warty cap, a wide, flimsy ring on the upper stipe and a bulbous base covered by a thin sheath. The sheath (or volva, as it is called) and the white patches on the cap are remnants of a "universal veil," a thin membrane that covers the entire developing *Amanita* when it is young. An *Amanita muscaria* in the immature button stage may resemble a puffball, until it is sliced open to reveal the silhouette of the future fungus.

The gills of a fly agaric are pure white, closely crowded, and considered "free," as they are not attached to the white or yellowish-white stipe. The *Amanita muscaria* found along the east coast of North America is typically the yellow and yellowish-orange variety, known as the *formosa* variation. On the west coast, the full range of colors of this species can be found, including the bright red fly agarics that are among the most photogenic and iconic of woodland mushrooms.

Notes: I noticed an *Amanita muscaria* on the cover of a recent issue of a trendy urban weekly, for an article on therapeutic hallucinogens. Despite the big red and white symbol of psychedelia on the cover, the article explored recent research on the effects of taking *Psilocybe* species (magic mushrooms), plants like peyote, and chemicals like ketamine and LSD, without ever once mentioning the fly agaric or its uses at all. Even though there is evidence that Indigenous peoples in Siberia and Scandinavia used these psychoactive mushrooms in religious rituals, and there are people around the world who have used and still use these specific *Amanitas* as a recreational intoxicant, most serious reference sources warn against the practice.

Popular culture is saturated with stylized (read: oddly proportioned and rarely morphologically accurate) images of these striking fungi, though, since they are so beautiful. In addition to their role as representatives of tripping on 'shrooms and psychedelic music, they can be found on Christmas ornaments and earrings and children's toys, and as symbols in the Super Mario video games, among so many other examples. The polka-dotted toadstool featured in illustrations of the pipe-smoking caterpillar from Lewis Carroll's *Alice in Wonderland* is usually a

Fly agaric,
Amanita muscaria

Fly agaric,
Amanita muscaria, var. *formosa*

red *Amanita muscaria*. Even the figure of Santa Claus himself, in his red and white suit flanked by reindeer, may have emerged from a creative hallucination among tripping Scandinavians, caused by the muscimol and ibotenic acid toxins found in these mushrooms.

The curious common name fly agaric refers to a historical practice of putting pieces of this mushroom into a bowl of milk to rid one's house of flies. The chemicals in the mushroom are said to both attract and kill the insects who dive in for a sip. I have considered checking to see if this gruesome prospect is true when the flies are buzzing monotonously and banging into the windowpanes at the end of summer, but I haven't tested it yet.

Edibility: *Amanita muscaria* is a poisonous mushroom and not safe for eating in any quantity. This species contains psychoactive and neurotoxic chemicals and is known to cause hallucinations, including confused impressions of scale, recalling Alice's sense of getting taller and smaller in Wonderland. Consuming fly agarics has also been known to cause tingling lips and tongue, euphoria, inebriation, heightened senses, dizziness, seizures, and vomiting, among other uncomfortable effects. In Lawrence Millman's test on himself, he describes hearing mushrooms talking and being able to communicate with a bottle of beer—even after he parboiled the fungus in an effort to reduce its toxicity. In another account, Eugenia Bone says that after her own experiment in tasting some, she immediately fell into "the kind of sleep you go into before a colonoscopy" and woke up wearing only one shoe.

While reports of death attributed to *Amanita muscaria* are rare, it must be considered that few people who have tried to consume this mushroom once ever elect to do so again.

Costume design by E. B. Brooks for the film *Fly Amanita* by David Fenster

FUNGIFEST

Interdisciplinary exhibition by Mark Allen and David Fenster with Machine Project

FUNGIFEST, CURATED BY MARK ALLEN AND DAVID FENSTER, was a two-week-long event that the Machine Project artists and collaborators created for the Hammer Museum in 2010, whereby fungi seemed to have taken over the entire museum. Originally founded by Allen, Machine Project was a community-based collective and cultural space that brought together numerous artists along with techies, scientists, hobbyists, poets, and other creative practitioners working in Los Angeles from 2003 to 2018. Among other works that playfully disrupted the structures and institutions of high culture, they performed tiny concerts for individuals in gallery cloak rooms, hosted an underwater art exhibition in a Santa Monica swimming pool, and dressed hundreds of museum visitors in noisy bells to fill the building with the music of their movements.

FungiFest featured a mushroom identification table covered by local species, like the kind you would find set up at the end of any foray, except this one was in a museum among artworks under strict environmental controls. Along with local mycologists available to answer questions and provide IDs, one performer behind the table was wearing a white *Amanita muscaria* costume, with an annulus for a hood (the skirt-like ring on the stem of some mushrooms) and a giant red-and-white cap on her head. There were screenings of mushroom-themed films and AV projections of spores under the microscope. There were also the Mushroom Dancers, who, while wearing tan-and-yellow tights and poufy white berets, improvised live choreography all throughout the event. They interpreted the slow, flowing growth of fungi, tumbled around the rooms like spores and crawled across the museum carpet like twisting mycelia looking for nutrients. And to complete the festive atmosphere, chanterelle-infused vodka cocktails and black truffle gelato supplied by a city ice cream shop were served. There was even music made strictly for the enjoyment of mushrooms (not humans!) and a week-long slime mold race, held by sprinkling dried *Physarum polycephalum* on a 4-inch track made of agar, a kind of performance art (or maybe sport?) that pushes at all limits of the imagination.

In his introductory essay on the project, Jason Brown describes the ironic possibilities of finding nature in the densely populated sprawl that is Los Angeles. With the mushroom as an example, the artists collected and shared fungi in the museum to show just how many of nature's astonishments are around us even in the big city, and how diverse these possibilities might be.

The artists of FungiFest hoped, as I do, that if we fall in love with the smallest and most ubiquitous creatures, we might also leave room for nature (the mushroom as well as everything else) to flourish. And their work is a celebration in its own right of the pleasures of the body and of being connected to one another; it reminds us of our place in the marvelous ecology of beings. As Brown writes, "Let us rejoice in the humble mycelia, spreading beneath fields and forests, parks and parking lots, with their quiet surprises and subtle beauty. Let us delight in the alien logic of intricate, sprawling, hidden organisms, and the secret fabric woven among us."

Ghost Pipe

Monotropa uniflora

Ghost pipe,
Monotropa uniflora

This unusual plant, known variously as the **ghost pipe** or corpse plant, is often mistaken for a fungus. Even though *Monotropa uniflora* is evidently a flower—complete with pistil, stamen, leaves, and seeds in its modest, downward-facing blossoms—it isn't green anywhere; rather, it's so thoroughly and entirely white that close observers of the forest floor are not just spooked by it, they don't know where to place it in the familiar categories of things.

Ghost pipe is one of very few plants that contain no green coloring because it has no chlorophyll. It is an example of a rare myco-heterotroph—meaning it is a parasite that taps into the networks of nutrient exchange between fungi (in this case, from *Russula* species) and the trees they live in partnership with. In fact, research on this plant from just the last few decades contributed to our understanding of the vast collaborative systems that exist along mycelial networks, connecting plants with fungi, and plants with one another, as a way to share essential nutrients and communicate. It even inspired the now famous term coined by the English mycologist David Read: the Wood Wide Web.

In his book, *Entangled Life*, Merlin Sheldrake describes the awesome significance of the discovery that plants and fungi exist within complex interconnected networks, shattering previously held scientific notions of individuality, autonomy, and competition among species. To make this idea of an interconnected network of beings relatable, he says, "it is the difference between having twenty acquaintances, and having twenty acquaintances with whom one shares a circulatory system."

Ironically, another term for *Monotropa uniflora* is "Indian Pipe," which, like many plant names with "Indian" in their common name, was imposed by settlers, often with disregard for the names and the knowledge of species that were already present in sites of research and settlement. It might not be surprising, then, that it took so long for the settler/colonial environmental sciences to discover that all beings are interdependent—even though this idea is central to the epistemologies of the Indigenous peoples of Turtle Island and has informed sustainable ways for humans to thrive on the land for millennia. Plant ecologist and member of the Citizen Potawatomi Nation Robin Wall Kimmerer reminds us of this idea of connectedness and reciprocity when she describes how there is no individual interest in nature, and that "all of our flourishing is mutual."

Ghost pipe is uncommon in most places, and in some cases it is very rare. Take a photo, make a drawing, stop in front of a patch to reflect on the interdependency of all things, but do avoid picking—and thereby destroying—these luminous, extraordinary plants.

Giant Puffball

Calvatia gigantea

Giant puffball,
Calvatia gigantea

ID Features: *Calvatia gigantea* both looks and smells like a volleyball. It has white skin that darkens to beige with age, surrounding white spongy flesh swollen into blubbery heaps. The interior of the mushroom is pure white when young, turning brownish green as the spores mature inside. This fungus produces the largest fruiting body of any mushroom I know of in North America, sometimes reaching heights of 3 feet or more. When a large *gigantea* is found, it is an awesome event. It is carried to the foray table with fanfare and appreciation. Impossibly white and big and round, right from out of the dark—as if the full moon had fallen from the sky into the woods and someone had rescued it.

Notes: It is said that a **giant puffball** produces enough spores that if you made a new puffball from each one, you would have a chain of fungi that could wrap itself around the earth: seven trillion spores in a head-sized specimen, each one potentially forming not only one new puffball but potentially an entire new colony, fruiting in ever-widening fairy rings around grasslands and woods. If indeed each spore in one specimen produced a new mycelium, we would be drowning in bloated saprobes covering all of the land and oceans on earth.

Photographing a giant puffball next to a toddler is a little-discussed trope of amateur photographic practices. And while this staging is meant to indicate that the fungus is as big as the baby, it has the unfortunate effect of making equivalencies between baby and fungus, an eerie, boneless blob that is exciting mainly for its edibility. From the extensive research I have done online, this fungus can be the size of a kindergartener! Sometimes the child and the fungus are found lounging together on the couch. I have also seen *Calvatia gigantea* photographed like a fat belly in front of an adult who embraces it with pride, fake-straining to express its weight; or it may be held up by two people, showing its largest side to the camera. There are also more than a few images of giant puffballs dressed up in sunglasses.

The urban legend that claims you will grow a giant puffball in your lungs if you breathe in near a sporulating specimen is not true—perhaps with the exception of the severely immunocompromised. In fact, a healthy person is always breathing near a sporulating fungus. We are inhaling spores now. At least two or three (or two or three thousand) spores from fungi of all kinds are around us and inside us, every single moment.

Edibility: *Calvatia gigantea* is edible when correctly cooked. People get very excited about them, and I've seen giant puffballs carved into slabs and sold at farmers' markets. With apologies in advance to everyone who loves them, I admit I have never once enjoyed cooking or eating giant puffballs. I find they absorb huge volumes of oil, and despite the promises and advice I have received on the matter, their flavor cannot be concealed by bread crumbs, herbs, cheese, garlic, or salt. They continue to look and smell like a volleyball and taste like dishwashing liquid. I might add another warning: If you forget or delay cooking your specimen right away, the clean white orb will incubate all its resident larvae, and you will wake up to a soggy mass of stuff writhing with maggots on your kitchen countertop.

Golden Spindles

Clavulinopsis fusiformis

ID Features: Golden spindles are found growing in a clustered bouquet on the forest floor or in the grass. They are bright yellow and wormy-shaped, with pointed tips. They should not be confused with what are officially known as club fungi, which are another genus of wormy mushrooms that flare or bulge toward the tips.

Notes: *Clava* in Latin refers to "club," but *-opsis* clarifies that they merely *look like* club fungi. *Fusiformis* refers to "spindles"— the smooth, pointed wood sticks used to spin wool into yarn. Like many mushroom names, it refers to old technologies that may need some clarification for modern mushroomers. Other common names for golden spindles are tongues of flame and slender golden fingers. My favorite contemporary appellation, invented by a keen eight-year-old mushroomer I met in the woods among these fungi erupting all over on Cape Breton Island, is french fry mushrooms.

Golden spindles,
Clavulinopsis fusiformis

Edibility: Despite their cheerful color and easy likeness to French fries, golden spindles are not considered edible according to North American field guides. However, food traditions vary around the world, and I have read that they are sometimes foraged and eaten in Nepal.

Green-Capped Jelly Babies

Leotia viscosa

ID Features: These adorable fungi are well named; they are short and clumsy-looking, with bright yellow stalks topped with dimpled green caps. They are found growing on wet ground in the woods or on mossy decomposing stumps, often in groups.

Notes: The **green-capped jelly baby** and its cousin, the all-yellow common jelly baby (*Leotia lubrica*), are rubbery and gelatinous, though they are not technically jelly fungi. They are considered sac fungi, meaning they have no gills— rather, they form spores in sacs within the fruit body of the mushroom.

Green-capped jelly babies, *Leotia viscosa*

Edibility: *Leotia viscosa* is not considered edible, though it is a notable little fungus to pat on the head and coo to in a high-pitched voice, if you find one in the woods.

Hedgehog Mushroom

Hydnum repandum

ID Features: The most distinctive features of **hedgehog mushrooms** are the spines under the cap, instead of gills or pores. These are members of a group commonly called the tooth fungi—though the words "spines" and "teeth" fail to conjure the masses of fragile, thin, downward-pointing icicle-shaped features that are the spore-producing parts of these mushrooms. *Hydnum repandum* are creamy to pale apricot-colored and will bruise light brown when damaged. Sometimes the mushroom has a fit and tidy symmetrical appearance with a round cap, and other times the stipe can be short and stocky and the cap thick and wavy around its edges. They grow in the woods on the ground.

Notes: Because of their creamy color and shape, hedgehogs may be mistaken in the field for faded chanterelle mushrooms, though one look under the cap should clear up this misidentification. Like chanterelles, their texture is suede-like and dry on the surface, and they possess a fainter version of the chanterelle smell that is sweet and appetizing. I have seen *Hydnum repandum* mislabeled in grocery stores as chanterelles; not only was this surprising, it made me wonder about the care with which wild mushrooms are checked, identified, and confirmed to be safe for sale. Though I admit, weary grocery staff have been unimpressed when I informed them of my urgent mycological insights. While the misidentification and mislabeling of fungi—by accident or for profit—is known wherever wild mushrooms are sold, hedgehogs are safe and delicious by any name and should be purchased and enjoyed with enthusiasm.

Edibility: Hedgehogs are considered edible when correctly cooked. They are considered among the easy mushrooms for new foragers to identify, because once you learn to recognize the spines and are confident you found these cap and stipe mushrooms growing on the ground (as opposed to growing from a tree), there are no known poisonous close look-alikes. They are not as common as chanterelles—or as fragrant, delicious, or colorful—but they are good and especially endearing when they conjure the spiky little rodents they are named for. Do serve, and say aloud for me, "hedgehog pie."

Hedgehog mushroom,
Hydnum repandum

Honey Mushroom

Armillaria mellea

ID Features: Honey mushrooms are very common and grow in large clusters on trees. *Armillaria mellea* (only one species among several similar ones also known as honey mushrooms) have long fibrous stipes with darker brown coloring near the base and a thick scarf-like ring near the cap. Caps are pointed half-spheres when young and flat, wavy disks when mature. The caps are golden brown and covered with small, dark "scales" that are concentrated in a darker raised bump in the middle.

Notes: This fungus is a major parasite that causes the deaths of the trees it infects. It is so successful at spreading and surviving in forests that a closely related honey mushroom called *Armillaria solidipies*, found in the Blue Mountains of eastern Oregon, is considered to be the largest and oldest organism on earth—at more than 890 hectares and between 2,000 and 8,000 years old! There is a poem by Laura Kasischke called "The World's Largest Living Thing" about the honey mushroom that describes it like an organ in the earth. She calls it the "world's moldy heart" and "a gorgeous sprawling brain, dreaming / you & me."

Edibility: *Armillaria mellea* are considered edible only in small quantities when cooked. While honey mushrooms are sometimes eaten, and many people enjoy them, several sources suggest discarding the tough stipes, parboiling the caps, and disposing of the water before cooking them again. The Roger Phillips guide I rely on suggests they have been known to cause digestive illness and "should possibly be considered poisonous." Honey mushrooms can be very exciting to new mushroomers, as they appear in huge golden bouquets at the base of trees in autumn and have an appetizing-sounding name. Unfortunately, "honey" refers to the color and not the flavor of the mushroom, which in my experience is quite bitter and not worth the fuss of parboiling and cooking or the risk of eating.

And honey mushrooms are not recommended for collection by amateur foragers, as there are numerous species of brown mushrooms that are hard to distinguish, many of them seriously toxic. Be extremely careful, when foraging for honeys, not to accidentally collect a **deadly galerina** (*Galerina marginata*, page 125) in your basket. These common, very small brown mushrooms also grow on wood

Honey mushroom,
Armillaria mellea

and can often be found at the same place as honey mushrooms. They are distinct from honeys in that they tend to be much smaller, with a smooth cap, rusty-brown spores, and light brown gills, and may show a thin brown ring on their fibrous stipe. Deadly galerinas are extremely dangerous fungi, with the same amatoxins found in the **death cap** (*Amanita phalloides*, page 60) except in even higher concentrations. Always collect your specimens carefully and double check each mushroom later to make sure nothing so unwelcome has found its way into your basket.

Indigo Milk Cap

Lactarius indigo

ID Features: The color of the **indigo milk cap** is rare among mushrooms and all living things: a startling blue-jeans blue. These fungi have funnel-shaped caps with concentric rings and a depression in the middle. The cap, gills, and stipe are blue. There may also be darker blue spots all over the short stipe. Like other milk caps, the gills of *Lactarius indigo* will exude a latex when cut. The "milk," as it is called, is also pure blue.

Notes: The only time I have ever seen this marvelous mushroom was on a foray with the Toronto Mycological Society. One milk cap appeared on the foray table, mangled a bit, and faded to silvery blue with age. Mushroomers can be excitable about any number of indiscriminate, semi-rotted bits of fungus, but the reaction to this mushroom was unforgettable. Vito Testa lifted it up to show the crowd, and they started chanting, "We want to see the milk! We want to see the milk!" He obliged by dramatically lifting his knife and slicing the gills above his head for everyone to see. Even though the mushroom was a little dry, and its exterior hue fairly grey, it bled droplets of milk that were unmistakably bright, beautiful blue. The crowd went wild.

Edibility: *Lactarius indigo* is considered edible when cooked. While the mushroom is not common in eastern Canada where I forage, it is more common in the US and Mexico, where it is even sold in markets. While reports on its flavor vary from "superior" to "indistinct" to "too peppery to finish," I am a little wary of a mushroom—or any food—that turns dark greyish in the pan. Some have had success retaining the color of the blue milk caps with vinegar or lemon juice, and I propose using them in adventurous dishes for an uncanny visual effect.

Indigo milk cap,
Lactarius indigo

INFINITY BURIAL SUIT

Mushroom burial suit project by Jae Rhim Lee

MERICAN MUSHROOM GURU
Paul Stamets has famously
proposed that mushrooms
might save the world, due in
part to the ability of fast-growing mycelia
to break down and digest environmental
pollutants. In his own experiments, fungi
are shown to remanufacture substances,
including the hydrocarbons in an oil
spill, into carbohydrates—essentially,
into fruiting mushrooms. While some of
his claims are disputed by professional
mycologists, and large-scale trials and
replication studies still need support, he
calls fungi "vanguard species" that make
way for other creatures to flourish. He
proposes an irresistible idea: that a toxic
dump could be transformed into a thriving
green expanse of life.

Based on ongoing study and application
of myco-remediation by Stamets and
others, the Korean American artist Jae
Rhim Lee created the Mushroom Burial
Project. Standing in what she teasingly

calls her "ninja pajamas" in a TED Talk about her work, she describes the numerous environmental toxins that contaminate our bodies—including preservatives, pesticides, lead, and mercury. When we die, she notes, these pollutants (and more when we are preserved by conventional funeral practices using formaldehyde) return to the air, water, and earth. In her eccentric research, she collected her own hair and nail clippings and worked with familiar mushrooms—enoki, king oysters, and others—to study how they might consume parts of her, with the ultimate goal that they finally consume all of her after she is gone.

This work culminated in the creation of the Infinity Burial Suit, a stylish black outfit that covers the body from head to toe, embroidered with spore-infused yarn in a dendritic pattern. The suit, she proposes, will reduce the impact of our own deaths on the environment. Jae Rhim Lee has since created her own company, called Coeio, through which she makes the suit available commercially.

While more than a few people have volunteered their future corpses to the artist's research, I especially appreciate the conceptual aspects of Lee's work. The burial suit reminds us of our ongoing connectedness to all other beings in the wider environment and of our common fate: to die and dissolve back into the same earth-soup. Like the Buddhist maranasati meditation, Jae Rhim Lee's work prompts us to accept death and our own inevitable decomposition. And like such morbid ruminations do, it expands our awareness of—and gratefulness for—our buzzing, thriving aliveness right in this moment, before we, too, are food for the mushrooms.

King Bolete

Boletus edulis

ID Features: The young caps of *Boletus edulis* are rounded and light pinkish brown with a pale bloom, while more mature caps are a darker reddish brown and widen out, even to the size of dinner plates, when fully grown. The cap can also be slimy when the weather is wet. As a member of the bolete family, this terrestrial fungus does not have gills; rather, it has a spongy pore surface that starts out firm and white and gradually becomes softer, yellower, and eventually more olive green.

The stems of a **king bolete** are distinctive in that they are often thick relative to the cap and balloon out impressively toward the base. They are white when young, becoming more tan with age, and textured by a very fine white netting called reticulation. The reticulation is not always obvious, and it can take some practice to see it. The flesh of the mushroom is white and does not turn blue when bruised or cut. If a tiny piece of a *Boletus edulis* is nibbled, it is mild-tasting and not bitter. King boletes are found mostly among pines, fir, and spruce, and sometimes in deciduous woods across North America, though they are most plentiful (enough even to be collected commercially) in the west.

Notes: King boletes are immensely popular across Europe. The French call them *ceps,* the Germans *Steinpilz,* and in the UK they are called penny buns for their coppery caps. Italians—who are especially known for cooking with them—call them *porcini,* similar to *porcelini* for "piglets," though it isn't known if this is because of the pudgy, nearly spherical shape of a young king bolete or because of its rich and savory flavor. While chanterelles and morels are the stuff of infatuation, king boletes provoke a kind of ecstasy in cooks and mushroomers who adore them. Even the serious modern king of mycology himself, David Arora, is theatrical in his description of this species. He calls *Boletus edulis* "a consummate creation, the peerless epitome of earthbound substance" and declares mischievously that this king is "the one aristocrat the peasantry can eat!"

I include myself in this club of the worshipful, since I find the form, fragrance, and flavor of the *Boletus edulis* so delectable that it unlocks some mysterious center of yearning and happiness in the middle of my chest. These fungi are also the supreme embodiment of the gastronomy of Italy to me, and I have

King bolete,
Boletus edulis

two recollections that I count among my most vivid sense memories of food, and of mushrooms. One is of a deep terracotta crock of polenta suffused with mountain cheese and porcini, served to me in the Dolomites as a stunned Canadian teenager. More recently, when I was lucky to be on a patio in Venice, there was a pizza *I never even ate*, piled with porcini and prosciutto and *tartufi* (black truffles), which was being carried—redolent and steaming—to fortunate strangers at the table just beside me.

So much do I love these fungi that I have stood enraptured, probably blushing, under the supermarket lights on the rare occasion of a display of fresh ones. I would describe the flavor of *Boletus edulis* to be like roasted meat and hazelnuts, mingled with the sweat of a man one loves entirely. And dried porcini—available at my local Polish market sewn into long, covetable garlands—are sour- and salty-smelling, like dried brine left on the skin of someone who has been swimming all day in the sea.

Edibility: *Boletus edulis* is considered to be an excellent edible when cooked. In coniferous woods in spring (depending where you are) and especially in the summer and fall, look for young specimens just pushing up through the leaf litter, or mature ones before the insects feast and deposit their brood all over them. This can be a serious contest, and sometimes, by the time you see large king boletes, they are tragically mushy and writhing with maggots and slugs. For some Italians, the quest to pluck a perfect porcino has sent them falling off the mountainside. Apparently, more people are killed this way every year in Italy than from mushroom poisonings! If you do find some great or at least salvageable specimens (on the safe side of the mountain), peel off any dirt from the stem with your knife and pile the clean mushrooms in your basket.

When sliced and cooked fresh, or from rehydrated or defrosted specimens, *Boletus edulis* are beloved for pickling; stuffing into ravioli, crêpes, or pierogi; topping polenta and pappardelle; stewing with game birds or venison for a ragout; and cooking with other autumn favorites like squash and chestnuts. Open a bottle of wine, toast to the season and love and the earth, and enjoy a glorious dinner.

Unfortunately, king boletes are elusive where I look for them in Ontario and the east coast, and I have a knack for finding a cruel look-alike, *Tylopilus felleus*, also called the **bitter bolete** (page 109), which has a paler cap and pores, and brown reticulation instead of white. When nibbled, this mushroom is awful-tasting and unfit for the pot. Other look-alikes

Bitter bolete,
Tylopilus felleus

include poisonous boletes, some of which have red pores or have flesh that turns blue upon slicing or bruising. The photogenic **cornflower bolete** (*Gyroporus cyanescens*, page 111) is among the few blue-bruising boletes that may be edible; however, it can be unappetizing as it darkens in the pan, and since there are toxic species that share this striking visible feature, it is worth avoiding the entire category of bluing boletes.

Depending on experience (and recklessness brought on by enthusiasm), one could even mistake a young, copper-capped, toxic *Cortinarius* species for a *Boletus edulis*. So, as with all mushrooming for the table, it is best to do your research and go out with more experienced foragers for a few seasons until you can identify *Boletus edulis* and a few other of its closely related and delicious cousins.

Cornflower bolete,
Gyroporus cyanescens

Lemon Drops

Bisporella citrina

ID Features: Lemon drops are very small, sturdy cups that are bright yellow on both the inside and outside. They appear in large groups on dark rotting logs, twinkling all over like fairy lights. They should not be confused with yellow, gelatinous, irregularly shaped witch's butter, which also fruits on decaying wood.

Notes: While their smell is not distinctive, you have to think of tiny candy orbs when you encounter them, and just imagine the perfume of lemons and sweetness in their proximity. When you find a spray of these bright yellow fungi in an otherwise grey and barren woods, it is a quiet joy. I recommend that you don't crudely scrape them into your basket, but if it's necessary to remove them, carry the stick or section of bark that they are fruiting on right out of the woods for identification and admiration.

Edibility: *Bisporella citrina*, unlike the old-fashioned sweets they are named for, are not considered edible.

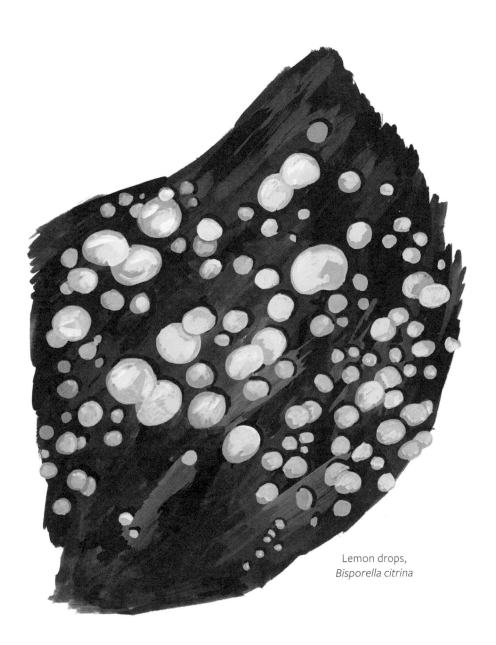

Lemon drops,
Bisporella citrina

Liberty Cap
(Magic Mushroom)

Psilocybe semilanceata

ID Features: With a top like a steeply pointed nipple, the shape of *Psilocybe semilanceata* has been compared to the knobby hat given to slaves in ancient Rome upon being freed—hence its nickname, the **liberty cap**. The cap of this mushroom is usually tan to chestnut brown and very slimy, and it lightens considerably when dry. Its gills are pale at first and become dark purplish-brown upon maturity. It has a long slender stalk marked with fibers and with a flimsy ring sometimes still visible near the top. Specimens are usually dirtied by smudges of dark falling spores and, distinctively, have flesh that turns bluish when handled or dried. (See **little brown mushrooms** on page 122 for more detail on the many similar, sometimes deadly, specimens, and more resources for serious foragers.) This very small mushroom grows in groups in grasslands across North America, especially in sheep and cow pastures.

Notes: Since the 1950s, when Gordon and Valentina Wasson wrote an article in *LIFE* magazine describing the Mexican mushrooms and rituals of shamans like the Oaxacan Maria Sabina, fungus species like the liberty cap have become notorious among mushroomers across generations. Magic mushrooms, or 'shrooms, are fungi containing psilocybin and psilocin, chemicals similar to LSD, which cause hallucinations and spiritual experiences when ingested. While the slender, dainty liberty caps are the most popular for foraging, especially in pasturelands and disturbed grounds in cities and gardens, most people in North America may be more familiar with the much larger, often cultivated variety, *Psilocybe cubensis,* known as the **golden top** (page 116). The golden top was championed by Terence McKenna, ethnobotanist and infamous psychonaut, who was a vocal proponent of exploring altered states of consciousness and inner worlds. He endorsed this species as part of his intentional movement to teach others how to grow their own hallucinogenic 'shrooms. He promoted the use of magic mushrooms as a form of mental resistance to mass cultural ideologies, suggesting that these make us slaves to consumerism and dangerously alienated from nature. He said:

This is a society, a world, a planet dying, because there is not enough consciousness, because there is not enough awareness. . . . And yet, we spend vast amounts of money stigmatizing people and substances that are part of this effort to expand consciousness, see things in different ways, unleash creativity. Isn't it perfectly clear that business as usual is a bullet through the head?

With a subversive counterculture growing around it in the 1960s and '70s, the suppression of magic mushroom use grew together with the popularity of the practice. As with many mind-altering drugs in North America, possession or sale of both species continues to be illegal in the US and Canada. In fact, there can be such a taboo around them that my strictly educational inquiries about identifying any magic mushrooms are met with squinty-eyed suspicion. Despite years of mushrooming, I can hardly find any experienced mycologists to talk to about them.

Until I began researching this book, I knew very little about recognizing and identifying psilocybin-containing species, and while I know they are hard to find, I am not even sure how often they show up on the ID tables. My impression is that senior mycologist notetakers are making discreet checkmarks on species lists and not saying a word about

Liberty cap (magic mushroom),
Psilocybe semilanceata

Golden top,
Psilocybe cubensis

them aloud. Paul Stamets calls out this profound reticence in his *Psilocybe* identification guide. He describes stoic, cautious mycologists who fear they might promote the harvest of illegal drugs, or worse, be liable for accidental poisonings. Less understandable are the conservative doctors, who Stamets says have pumped the stomachs of nervous trippers to "teach them a lesson."

Even Long Litt Woon, in her book *The Way Through the Woods,* about grieving her husband's death while discovering mushrooming in Norway, describes how hard it was for her to learn anything about magic mushrooms, in a chapter called "The Mushroom That Must Not Be Named." After futile attempts to glean information from a senior mycologist—who told her if she tried to take magic mushrooms she would end up in a coma—she held clandestine meetings in a café with an anonymous informant called "N" who described the effect of a mushroom trip as a beautiful thing and a way to see the world, they said, "without any filter . . . it doesn't provide you with any answers, but it helps you to see the world more clearly."

Edibility: *Psilocybe semilanceata* is toxic and is not recommended for eating. Possessing, growing, selling, or using this mushroom is also currently illegal, except in some regions for medicinal purposes. It is crucial to know that among the many little brown mushrooms that can be mistaken for magic mushrooms, there are some that can cause an agonizing death. As the expression goes, you absolutely *must* "know your 'shrooms."

Any serious foragers would be well advised to spend a considerable amount of time sharpening their mycological literacy and reading through Paul Stamets's highly regarded book on the subject, *Psilocybin Mushrooms of the World.* He talks about habitat, identification, dosages, who should *not* take magic mushrooms, and ritualized ways to lessen the risks these mushrooms may pose. Even when you get the identification of your magic mushroom right, symptoms of ingestion on the more extreme side may include terrifying hallucinations, panic attacks, exacerbation of mental illnesses and/or epilepsy, heart irregularities, allergic reactions, and a devastating loss of muscular control that can last for many hours or even days.

On the milder side of the experience, especially for trippers who are supported by trustworthy and experienced guides, people report visual and auditory hallucinations, euphoria, nausea, and vomiting—but also intense feelings of connectedness to nature, loss of ego, profound visions, and a strengthening of resolve to live with greater purpose and intention. While my own experience is limited to a single episode of nausea, teeth-clenching, and uncontrollable fits of laughter, artist friends of mine who use them regularly report expanded creativity, fresh perspectives on old ideas, and a delight in microdosing (taking very small amounts of mushroom tea or gummies) while they work privately in their studios. Other artist friends like to take mushrooms when camping, to roll around in the moss and feel themselves to be an ecstatic part of the universe. Paul Stamets himself says he goes out into the mountains to take them only once or twice a year, "to recalibrate my cerebral and spiritual compass." As an exceedingly cautious person myself, I admit I am envious of their private insights—and their abandon.

In addition to what are described as psychedelic, creative, and spiritual effects, several *Psilocybe* species have shown significant promise when used as treatments for depression, addictions, PTSD, OCD, and anxiety, and they are increasingly being authorized for use in research and therapeutic settings (as well as being used in all kinds of popular, though unregulated, settings). There is some hope they might be better studied and understood, and more widely available in conditions that do support more positive experiences. As a sign of the resurgence of interest in these fungi and rituals around their ingestion, Michael Pollan's book *How to Change Your Mind: What the New Science of Psychedelics Teaches Us About Consciousness, Dying, Addiction, Depression, and Transcendence* goes into the subject in great detail.

Lion's Mane

Hericium erinaceus and *H. americanum*

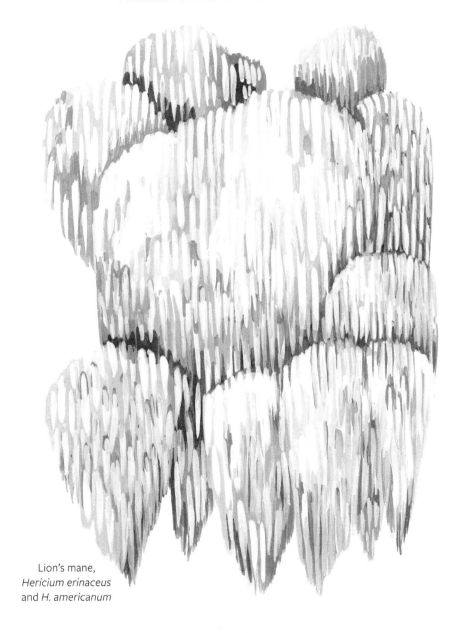

Lion's mane,
Hericium erinaceus
and *H. americanum*

ID Features: A **lion's mane** is among the easiest mushrooms for beginner foragers to identify. This fungus is distinguished by its coral-like branching structure, covered by downward-pointing spines, or "teeth." (True coral fungi have upward-pointing spines, like ocean coral.) Widespread and fairly common, these cream-colored fungi can be found growing like a miniature frozen waterfall from rotting logs or even jutting out of a hole carved by a woodpecker. They can grow to be larger than a porcupine.

Notes: While some look like a long-tasseled cheerleading pompom, my favorite *Hericium* species are the ones that resemble a starkly vertical mountain range from a seventeenth-century Chinese landscape painting. These are gorgeous mushrooms that challenge our mental picture of what a mushroom is.

Hericium species deserve admiration simply for being so strange and beautiful in the wild, but they are also popular for cooking, eating, and medicine. You may have heard of lion's mane as a popular commercial mushroom, as there is an incredible profusion of grow-your-own kits, capsules, tinctures, extracts, and even latte powder for sale. In fact, a quick search for "lion's mane" turns up so many mushroom-related articles and ads that I was surprised to see the ubiquity of these products completely overtake the subject of *actual* lions!

Nutraceutical companies do make a wide array of health claims about the lion's mane, including that taking it may treat ulcers, cancer, and heart disease; manage symptoms of diabetes; support the nervous system and brain function; heal wounds; "optimize" the immune system; encourage weight loss; and even protect against dementia, high cholesterol, depression, and anxiety—the latter being the only thing I'm convinced you could reasonably expect to reduce, if you believe in this easy cure for absolutely everything else. Despite the abundance of options and claims, and while there may be some promising indicators of benefits from taking supplements of lion's mane fungus in animal studies, even the National Center for Complementary and Integrative Health in the US warns that direct evidence of benefits to humans from consuming this mushroom for medicinal purposes is lacking. And, of course, it is also unknown whether taking the concentrated doses in these supplements, especially for an extended period of time, is in itself necessarily safe for everyone. While lion's mane is a fun and interesting edible mushroom, always use caution when trying new foods and taking anything in medicinal doses.

Edibility: *Hericium erinaceus* is considered edible when correctly cooked. Like many similar *Hericium* species known variously as coral tooth or bear's head, a lion's mane is a showy prize to find in the wild, and good for cooking. Look for firm, light-colored specimens (they turn mushy and brown as they age) and carefully clean them as well as you can before you put them into your basket, because later it can be tricky to remove dirt from among the thousands of fragile spines. I find a gentle soak in cold water will release some, in addition to flooding out some resident beetles. Lion's mane is lightly flavored, perfumy, and fresh, as opposed to savory and earthy. Many recipes suggest cooking and flavoring the unusual needle-shaped flesh like fish or crab. The last time I found one we cooked it among oyster mushrooms with garlic, ginger, soy, and sesame oil, and stuffed them into Chinese-style dumplings. They were delicious.

Little Brown Mushrooms

Various species

In popular parlance, little brown mushrooms (or LBMs) can sometimes refer to magic mushrooms (see **liberty cap**, *Psilocybe semilanceata*, on page 114). To mycologists and participants at forays with your local mycological society, however, this term is not a specific identification for anything at all.

Because there are thousands of small, brownish, and dull-colored mushrooms and encyclopedic volumes are filled with descriptions of them, it is very difficult to identify many of the odd little brown fungi you may find. The eccentric and colorful mushrooms are much easier to remember. With practice, though, you can learn a few of the more charismatic and common species of little brown fungi, and find some familiar creatures everywhere you go, no matter how dry or poor in mushrooms a site may be.

Outside your door right now, for example, you may find a **glistening mica cap** (*Coprinus micaceus*, page 123), a cousin to the tippler's bane and shaggy mane. These small mushrooms have tan and greyish bell-shaped caps that actually sparkle—like mica—in the sunlight. The thin flesh and gills on the cap will curl upward like a blown skirt when mature, before all the mica caps melt into a smudge of black inky goo in the grass or at the edge of stumps.

On a log, you might also encounter a lone tan **deer mushroom** (*Pluteus cervinus*, page 124), distinctive for its pale gills and salmon-pink spore print. Or, among the most photogenic of little brown mushrooms are *Marasmius* species, including *Marasmius rotula*, an exquisite specimen found growing in impressive quantities all together on their tough, long skinny stems in troupes. *Marasmius rotula*'s distinctive cap is more beige than brown and seems carved with deep, radiating grooves—so it is sometimes called the **pinwheel mushroom** (page 124). As long as you are not trying to eat what you find, getting a little brown mushroom into the right genus, or guessing at a relevant common name, is about as much work as you need to do to enjoy a mushroom foray in shades of brown.

If you are looking for liberty caps or other psilocybin-containing magic mushroom species, however, the misidentification of a little brown mushroom can have grave implications. For example, growing from your lawn right now, you might find a slender **haymaker's mushroom** (*Panaeolina foenisecii*, page 122), a toxic look-alike species to

Glistening mica cap,
Coprinus micaceus

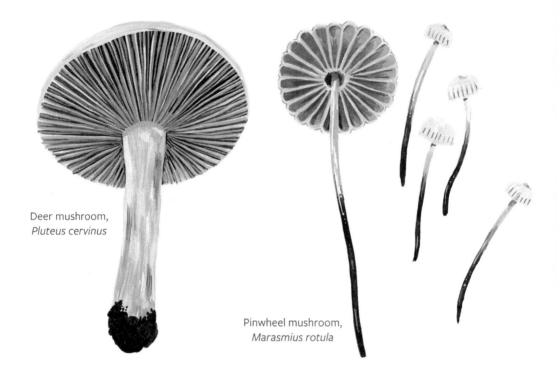

Deer mushroom,
Pluteus cervinus

Pinwheel mushroom,
Marasmius rotula

magic mushrooms, with bands of color in rings around its cap. This is a common mushroom growing in grassy areas, with purple-brown spores and sometimes a dark brown nipple-like point on its beige cap, that will definitely make someone sick if ingested.

For foragers of magic mushrooms, an even worse misidentification would be for a **deadly galerina** (*Galerina marginata*, page 125) and many of its similar-looking, seriously poisonous cousins. Also

sometimes confused for psilocybin-containing 'shrooms is the **deadly conocybe** (*Conocybe filaris*, page 125). This dainty pointed fungus has ridges on its cap and a fluffy ring around its slender stipe. Despite its graceful appearance, it is as poisonous as a **death cap** (page 60), causing multiple organ failure and an excruciating death.

So when a foray leader declares something to be a little brown mushroom, they mean it is probably too small, too

Deadly galerina,
Galerina marginata

Haymaker's mushroom,
Panaeolina foenisecii

Deadly conocybe,
Conocybe filaris

inedible, or too boring to spend much time fussing over, and that it could just as easily be a magic mushroom as one of the numerous little brown fungi that could kill you. A long apprenticeship with more experienced foragers, deepening your mycological literacy, and careful study of books like Paul Stamets's identification guide to *Psilocybe* mushrooms may be in order if you really want to know the nature and identity of any one species.

The truth is, though, most little brown mushrooms are impossible for foragers to identify with certainty. So if you are looking for 'shrooms to trip on, you should summon your humility and acknowledge the limits of your identification powers. It takes time—and experience. If you aren't looking for 'shrooms, you will find it's nice to just be curious, appreciate things as they are, and accept more of the woods' many possible mysteries.

Lobster Mushroom

Hypomyces lactifluorum

ID Features: A **lobster mushroom** is flame red on the exterior and can be so saturated and vivid that it draws gasps. Theoretically this mushroom is shaped like a thick-bodied flaring funnel on a short, squat stipe. But an actual specimen is often deformed and curled into itself, bringing dirt and pine needles and resident insects (and sometimes neighboring lobster fungi) along with it. Its surface is like a thin, grainy crust that is almost glassy and crystalline. And like a cooked crustacean, a lobster mushroom is bright orange-red on the outside and pure white on the firm, resilient inside.

Notes: The lobster mushroom is an excellent example of a fungus on a fungus: it consists of a thin colony of *Hypomyces* fungi parasitizing another mushroom, usually a brittlegill (*Russula*) species or a milk mushroom (*Lactarius*) species. The action of the parasite profoundly changes the form and surface of the host, creating a hybrid that is both dazzling and monstrous.

Speaking of monstrous, a similar relationship results in a fungus with the unfortunate common name **aborted entoloma** (*Entoloma abortivum*, page 129).

It consists of a purple-spored *Entoloma* species parasitized by a honey mushroom (*Armillaria*) species, or possibly vice versa. Mycologists are often too distracted by the convoluted, gill-less flesh and spongy skin of this less popular edible mushroom to be sure. The entire situation of a handful of misshapen "aborted" fungi on the foray ID table can provoke a moment of very uncomfortable silence for leaders and participants, especially after the name is announced. For people who do eat these fungi—usually battered and fried—they are considered excellent, though, and sometimes given the more appetizing name "shrimp of the woods."

Edibility: *Hypomyces lactifluorum* is edible when cooked but may not be tolerated by everyone. You will need to try a small quantity at first, with caution. This is a complicated mushroom, since the parasitic fungi that make a lobster a lobster disfigure their host so much that they make it impossible to identify the host species and evaluate the edibility or toxicity of the *Lactarius* or *Russula* species inside. This fungus has been eaten by many people for a long time, though; it is offered for sale in markets and stirred

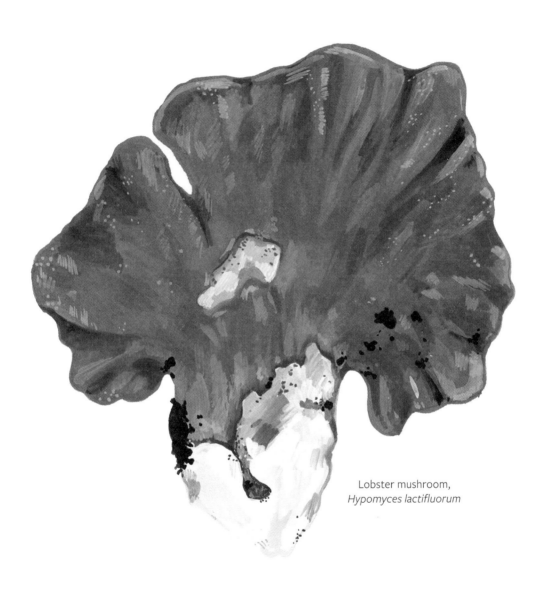

Lobster mushroom,
Hypomyces lactifluorum

into expensive pasta dishes, and I have not encountered any reports of resulting poisonings. It is not clear whether the *Hypomyces* parasitizes only edible host species or changes the nature of any toxicity in the host, possibly making an inedible mushroom into an edible one.

When eaten, a lobster mushroom is considered choice, and it will bring flashes of color to a dish with a mix of wild mushrooms. I consider it a treasure to find a sizable lobster mushroom, as they don't appear often or in large groups in my region. Lobster mushrooms are comparable to the ocean-dwelling crustaceans they are named for in both taste and luxuriousness, and they are often cooked with seafood (or used as a vegetarian substitute). I have found my own lobster mushrooms to be somewhat fishy-smelling, though I'm not sure if it's because they were decomposing a little or because they look so much like lobster—even chopped and sautéed—that I experience them as distinctly briny.

Aborted entoloma,
Entoloma abortivum

Matsutake

Tricholoma magnivelare and *T. murrillianum*

ID Features: The Japanese word **matsutake** translates to "pine mushroom," as these mushrooms are found in symbiotic association with pine trees (and sometimes fir), mainly in regions of the Pacific Northwest. These are medium to large mushrooms that are mostly white, with brownish fibrous scales on the cap. *Tricholoma magnivelare* have numerous closely spaced white gills and make a white spore print. When they are in the button stage, a thin white membrane covers the gills. When they are mature, the cap flares out to a large umbrella and the membrane collapses into a fragile ring, if still visible at all. Above the ring line, a matsutake is white, while underneath, it is usually scaly and brownish and sprinkled with the sandy to ashy soil and conifer needles that are present where it fruits. The flesh of the matsutake is very firm, like a strong rubber, especially when young.

These fungi are also identified by their smell—something an eighth-century Japanese poem describes as the "aroma of autumn." Modern North American guidebooks describe it to be like cinnamon, or sometimes less generously like "stinky socks," though I disagree. Despite how incredibly prized matsutake are in Asia,

Europeans have famously eschewed this fungus and have labeled their own similarly odorous species as *Tricholoma nauseosum*. I would describe the aroma as pleasant, almost synthetic—like cinnamon gum or a spicy perfumed soap.

When you go to look for them, keep an eye out for a lip of white with some rusty brown marks peeping out from under the green moss. It is easy to spot large, fully emerged specimens, but the younger and most prized pine mushrooms are dug up from under a notorious "shrump"— or mushroom bump—that is pushed up by the tough mushrooms growing underneath. Check by look and feel within a few feet around any visible specimens, for the treasure of some partly buried neighbors.

Notes: Japanese immigrants to North America in the nineteenth century initiated the hunt for this familiar *Tricholoma* species while they worked as migrant laborers clearing forests and building railways in the US and Canada. During World War II, when Japanese families were removed from their homes and forced to live in isolated, brutal internment camps, many had their property and

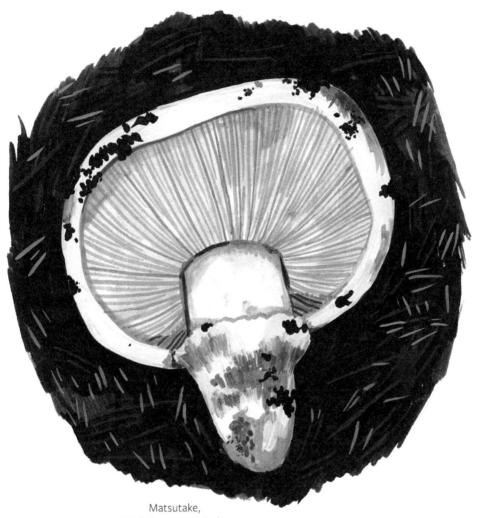

Matsutake,
Tricholoma magnivelare
and *T. murrillianum*

livelihoods stolen in payment for their own imprisonment. After the war and to this day, Japanese people in the Pacific Northwest hunt for matsutake together with their families as a way to practice their inherited culture and to make meaningful connections with the land they were once exiled from.

In addition to Japanese families, there are numerous communities of matsutake pickers in the west, particularly Indigenous people, Vietnam War veterans, and migrant workers—all enduring difficult living conditions in the forests to supply a competitive commercial trade in the valuable fungi. In recent years, matsutake considered to be of the highest quality (young, firm, and with the veil still covering the gills and looking unmistakably phallic) have sold for up to a thousand dollars per kilogram in Japan.

In her recent book, *The Mushroom at the End of the World*, Anna Lowenhaupt Tsing uses the example of the matsutake to describe how a species that thrives in ruined environments (often in logged woods) shows how resilient natural systems are, and how economic opportunities are seized and self-organized in response to a surge of material abundance. In this beloved cross-disciplinary publication—often quoted by my most earnest, anti-capitalist graduate students—Tsing calls for an acknowledgment of the most precarious members of our interconnected systems— plant, fungi, animal, and human—and for mindful, collaborative solutions to the project of our mutual survival. As a subject of centuries-old Japanese poetry, a story of the movement across the world of migrant workers, a reminder of devastating xenophobic national policies during wartime, a case study in the vulnerability of Indigenous peoples and veterans, a model for economic and environmental resilience, and, finally, an expression of the highest levels of gastronomic pleasure and excess, the matsutake is an excellent example of something one might think of as natural—just a *fungus*—but that is, rather, a distinctly cultural artifact. It is the story we tell about it.

Edibility: *Tricholoma magnivelare* and *T. murillianum* are considered edible when correctly cooked. Matsutake are not easy mushrooms to identify, and there are deadly poisonous look-alikes, including various *Amanita* species. While *Amanitas* have a distinctly swelling or bulbous base (*Tricholoma magnivelare* tends to have a more tapered base) and are a little more fragile-fleshed, they can both be white-gilled, white-spored white mushrooms with a ring, and very hard for anyone but experienced mushroomers to distinguish with confidence. I strongly recommend using multiple guidebooks and, if possible, practicing finding and identifying matsutake with experienced foragers. It is rare, but if you do find a person big-hearted enough to share their pine mushroom spots, know that you are receiving a munificent gift.

Anna, my friend and generous pine mushrooming mentor, finds them in great numbers around her home in coastal British Columbia. She told me that she once found so many that she had to process (brush, wash) them all in her bathtub! She likes to grill thick slabs of the pines with a little olive oil, salt, and lemon, and freeze the rest for later, whole. I do find freezing them affects their texture and their delicate fragrance, though, so they are optimally eaten in season, and fresh.

If you are lucky enough to find matsutake, either in the woods or at your local Japanese specialty shop, one or two precious specimens sliced thinly go a long way for flavor and aroma. My favorite way to prepare such a precious sampling is to bake the thinly sliced caps in a little foil pouch with sesame oil and whatever Japanese ingredients I have on hand— including ginger, soy sauce, and mirin. And instead of wasting any tough fibrous stems of larger specimens, I put them into my rice cooker, to perfume some sushi rice while everything steams together. A spoonful of mushrooms and their cooking juices in a bowl of scented rice is fresh and fragrant, and should be enjoyed like a poem about autumn, written by the earth.

Platter featuring numerous reptiles, amphibians, small sea creatures, and insects, made in France by a sixteenth-century follower of Bernard Palissy, from the permanent collection of the Metropolitan Museum of Art

THE METROPOLITAN MUSEUM FORAY

Museum-based foraging project by Diane Borsato

As an artist and art educator, I have always been interested in museum collections and the stories we tell about the objects in them. As a naturalist, I have discovered that the methods of observation I have practiced in the field have given me a unique set of tools and vocabulary for seeing artworks and artifacts more closely, and for interpreting them in ways that expand beyond the familiar art historical, nationalistic, colonial, or commercial frameworks.

As an official appreciator of clouds (I'm member #12,671 of the Cloud Appreciation Society, which is a real thing!), I noticed I could identify all sorts of weather and optical phenomena in historical Canadian paintings and notice how many sunny days were depicted compared with how very few representations there were of *rain* in Canada. It turns out that if you look at a national collection of landscape paintings this way, you not only find that there is no unsightly weather in the public history, but might also notice that many other unflattering truths—things like so much colonial violence on the land—are conspicuously absent, too.

When I read about the American artist and avid bird-watcher Mark Dion actually going *birding* at the Metropolitan Museum of Art in New York, I learned he and his companions identified an impressive number of species in the galleries, including magpies in a Goya painting, mute swans swimming around on the Tiffany glass, and vultures posing ominously on Egyptian urns. It seemed inevitable that if there was this much avian wildlife among works in the collection, I might find fungi there, too. And I wondered what else I might find if I tried an entirely new way of being, and seeing—with the museum as my wilderness.

In February 2020, I put on my most comfortable walking shoes and set out with the goal of doing research for a project called *The Metropolitan Museum Foray*, determined to find all the mushrooms that I could in the museum. I walked for seven and a half hours that day, in and out of every single gallery in that enormous institution, with my best mushroomer's eyes—the kind I use to find the most camouflaged morels among the leaf litter—screwed into my skull.

I noticed scarabs, dragonflies, and hummingbirds. I saw cats, rats, horses, cows, deer, dogs, plenty of wolves, bears, foxes, and rabbits. I think I saw every kind of flower and every kind of leaf. I saw snakes, lots of them—a shocking number of which were strangling figures of women. I was looking so hard for mushroom-shaped silhouettes or forms of any kind—at tapestries and statues and screens and paintings and ceramics and jewels and tools—that a lens actually popped out of my glasses! I saw annunciations and crucifixions, every form of battle, and every form of weapon. I saw all the paintings and sculptures of nursing mothers holding their babies, which I love as a category of art from across geography and time—though it occurred to me that there were so many more depictions of rape throughout the museum than expressions like these, of human tenderness.

While I saw mushroom-shaped hats, and cup-like cups, and club-like clubs, and one possible lichen—it was all a stretch to consider them fungi, even for my eager imagination. In the end, after the closest, hardest looking I have ever performed, I could say there were no intentional depictions of recognizable fungi, *not one single mushroom anywhere in the entire Metropolitan Museum of Art.* I was overstimulated, exhausted, and bewildered. I even felt betrayed by the place, which seems to promise everything on earth inside it. How could this whole kingdom of life be missing?

Since my first foray at the Met, I've since learned there are at least a few representations of fungi deep in the storage vaults of the museum. There is an eighteenth-century jade sculpture from China of a healing lingzhi, and an endearing wooden netsuke from nineteenth-century Japan shaped as a mouse atop a pile of mushrooms in a little basket. I would argue, too, that the graphic screen print of the 1968 *Cream of Mushroom from Campbell's Soup I* by Andy Warhol counts as another

Netsuke of *Mouse in a Basket of Mushrooms*, Japan, nineteenth century, from the permanent collection (not currently on display) of the Metropolitan Museum of Art

expression of the important cultural relationships human beings have made with members of Kingdom Fungi.

But just like in a day when I don't find mushrooms in the woods, I collected a basketful of other gifts at the Met foray. I have visited and strolled around many museums in my life, looking up and down at didactic panels and through the glass for my favorite kinds of things. But this effort to *be* and to *see* like a mushroomer in this context surprised me. It's not just the endurance I could summon, and an openness to surprise in dusty galleries I might never have been interested in before, but also just how alert I could be to the closest details in things, wriggling in the dark margins everywhere. And because I was scanning so broadly rather than getting lost in individual artworks, I could see the whole museum the way the poet Steven Wright describes the dictionary, like "a poem about everything."

✦ ✦ ✦

I'd like to go again for a foray in the galleries with members of the New York Mycological Society, if they will join me next time, to have more expert eyes scanning for mushrooms. We'll see what else—mycological and otherwise—we might find in the Met, in a different season, in different weather.

Mushroom Studio by Katie Bethune-Leamen, commissioned
by the Toronto Sculpture Garden in 2008

MUSHROOM STUDIO

Public sculpture by Katie Bethune-Leamen

TORONTO ARTIST KATIE BETHUNE-LEAMEN describes her
enormous public mushroom sculpture as being like a whimsical
roadside attraction: something that is both amusing and,
inevitably, melancholy, as its handmade surfaces weather in the
sunlight. Sitting in a garden among shops, a church, and noisy streetcars,

the *Amanita pantherina* also has a door and windows in its giant stipe, opening into a small artist's studio inside. It's a beguiling vision, and Katie invited several local artists to do short, solitary residencies inside the mushroom, insulated against the world.

Katie herself spent a considerable amount of time there—the artist inside the art—working on drawings at a little drafting table under a reading lamp. When I saw this public sculpture, especially with an artist drinking tea and working inside it, I thought, too, about the historical relationship between mushrooms and drawing. Beatrix Potter is the most famous example of an illustrator who developed her technical skills by close observation of nature, particularly of fungi and lichen in the Scottish countryside and English Lake District. Including details of all their minute striations and scales and worm holes, their membranous rings, colored zones, and undulating caps, she made at least 350 accurate, charismatic drawings of mushrooms, for which she is especially known.

The giant storybook toadstool draws attention, too, to how (practically) impossible it is for artists to make a living in a big city and how fantastical it may be to be occupied by creative work at all. Are artists like mushrooms, too, because sometimes they make everyone uncomfortable—for their eccentricity, or sexuality, or their abandon of practical responsibilities? And sometimes, just like mushrooms, artists pop up in unlikely places and make unfamiliar and surprising new forms in the dark.

Mushroom Studio also makes me wonder if one of the reasons so many artists are obsessed with fungi is that they show us a more magical world that *is* real: another, more colorful realm to lose ourselves in and to feed our imaginations. The sculpture, like any still and absorbing attention to Kingdom Fungi, is a respite from the clattering of our too-real, Kingdom Animalia–based lives.

Old Man of the Woods

Strobilomyces floccopus

ID Features: These monstrous-looking fungi are considered boletes because they have a spongy, pore-filled surface, instead of gills, under their caps. The pore surface of a *Strobilomyces floccopus* stains black, and its flesh stains faintly reddish and then black when cut—not that you would need to see this, since the **old man of the woods** is an easy mushroom to recognize on first sight, by the black, hairy, pointed tufts all over its cap and stipe. The old man of the woods is found on the ground in the forest among oak trees, with which it shares nutrients underground as a mycorrhizal partner.

Notes: While it may not be well known in North America, the old man of the woods seems to be more popular and more admired in Europe. The government of Poland even made an attractive postage stamp of it in 1980, and the Swiss made one with this black, hairy little beast of a mushroom more recently, in 2014.

Since so many botanical and mycological specimens—and astronomical and zoological specimens, and so on, and so on—are named after the scientists who were the first credited for describing them, there are many more examples of mushrooms named after real old men of the woods. Among these gentlemen mushrooms is the striking **Frost's bolete** (*Butyriboletus frostii*, page 143), named for the amateur American mycologist Charles Christopher Frost, which looks like it was chiseled roughly out of bright red wood. *Caloboletus marshii* is named for its discoverer, Ben Marsh, who, according to David Arora was as "dense, bulky, bitter, and bulbous" as the inedible mushroom named after him. Other masculine characters like the **king bolete** (*Boletus edulis*, page 107) himself, and a distinctively ringed, intensely slimy fungus known as **slippery jack** (*Suillus luteus*, page 142), are also in their company.

Edibility: *Strobilomyces floccopus* is considered edible when correctly cooked. This mushroom, along with a few extremely close species including the equally hairy *Strobilomyces strobilaceus*, is considered fairly easy to identify, and without poisonous look-alikes. Old man of the woods is therefore often recommended as an edible for new foragers to learn, but unfortunately the flesh of this mushroom turns an

Old man of the woods,
Strobilomyces floccopus

Slippery jack,
Suillus luteus

unappetizing black in the pan and it is not known to taste very good. People do eat it, however, especially if they follow all of the advice to make it more tender and flavorful. But gastronomes be warned:

Michael Kuo describes what is left of this mushroom after you wash off the tufts, remove the pores, and discard the tough stipe as a "mysterious looking morsel resembling discolored tofu."

Frost's bolete,
Butyriboletus frostii

Orange Birch Bolete

Leccinum versipelle

ID Features: Orange birch boletes are found growing on the ground in association with birch trees in the late summer to fall and are widespread across North America and Europe. They are part of the bolete family, as they are stalked mushrooms with spongy pores under their tawny orange caps, instead of gills. The flesh of the mushrooms turns purply grey or black when cut. Boletes in the *Leccinum* genus are distinctive in that they have black scaly markings all along their stalks. The best comparative I have heard for these scales, or more properly scabers, is that they look like the dark stubble on the face of a man who hasn't shaved in a week.

Notes: If you catch them at the right moment of size and freshness, these mushrooms are very good-looking, with their orange caps and conspicuous black markings against a tall white stalk. When I think of them, I see a woodland full of pale young birches, all their leaves shimmering in the sunlight. No matter how long I've been mushrooming, though, I can't say I'm confident distinguishing these boletes from their many confusingly similar *Leccinum* cousins, including the orange oak bolete, the orange bolete, the dark-stalked bolete . . . and so on. Some focused research has just made the situation worse, as I've read of caps changing colors with maturity and *Leccinum* flesh that turns various colors when cut—pink and partly green and then black, etc.—and how many similar names there are for both similar and dissimilar species, and I have thrown my hands up in submission. I witnessed this coloration drama when I tried to identify handsome **brown birch boletes** (*Leccinum scabrum*, page 146), with a dry, silvery-brown cap that doesn't bruise at first but turns quietly pink and then slowly bluish teal on the stems and then greyish black. They are all just *Leccinum* to me now, and I greet them humbly, whichever species they might be.

Edibility: *Leccinum versipelle* is technically a toxic mushroom, but it is considered edible when properly, *thoroughly* cooked. This may sound confusing and unfortunate, as these are attractive-looking common fungi, and they are indeed collected for the table in Finland and many other European countries. I have read of several accounts of unpleasant reactions to this mushroom, though, including gastrointestinal distress and vomiting. The

Orange birch bolete,
Leccinum versipelle

Brown birch bolete,
Leccinum scabrum

instructions for preparing this mushroom to eat include boiling or frying it for more than 20 minutes to reduce the toxicity. While I would hazard a taste of this mushroom if it were offered to me by an experienced forager and knowledgeable cook in a place where these are commonly collected and known, I cannot confirm whether this cooking method makes orange birch boletes safe, and wouldn't personally bother boiling and frying it to test it out on myself. I have collected a few brown birch boletes (which are considered safer to eat, but with caution) and sliced them thinly before roasting thoroughly with sesame oil and salt. The little *Boletus*-flavored slivers were delicious with rice, and I felt no ill effects.

Oyster Mushroom

Pleurotus ostreatus

ID Features: Oyster mushrooms are mainly found growing on rotting logs or dead trees from spring to early winter across North America. These common fungi often appear in groups, sometimes emerging from the same base in a bouquet, shelving attractively in clumps. They have a short white stipe, with white- to cream-colored gills that run without stopping up from the stipe to the underside of the fan-shaped beige to tan to mauve cap. The spores of oyster mushrooms are white. When fresh, they have a mild and pleasant smell, which could be described as faintly oystery or oceanic—but this might be the suggestive effect of their marine-related name.

Notes: Among the most surprising and persistent trends that arose during the COVID-19 pandemic, along with baking sourdough bread, was growing fungi at home in DIY grow kits. The kits consist of sawdust or other substrate material that has been injected with spores of edible wood-decomposing fungi, including oyster mushroom varieties as well as **lion's mane** (*Hericium erinaceus*, page 119) and **enoki** (*Flammulina velutipes*, also known as **velvet foot**, page 50). After you soak and spray the unpromising-looking bag of sawmill waste, the fungal spores grow into thread-like mycelium and quickly devour the food. They burst into fruit through cut-outs in the bag and, in the case of oysters, expand into curvy shelving fans at startling speeds. They grow noticeably faster than plants do, and watching them swell gives you the uncanny feeling you are seeing time-lapse film play out live. And since fungi are so adaptable and sporulate in such aggressive numbers, spores do escape into the air and can develop elsewhere in your house, including in your leftover coffee grounds or in your countertop compost. My partner has grown many kits in our basement over the years, and after the first picturesque fruiting (and perhaps some over-eager spraying), the kits become host to other, uninvited spores present in the environment and become soupy and furry with mold. *Pleurotus ostreatus* is also the species that is commonly available at supermarkets in North America, commercially grown in conditions that are necessarily more sterile.

Edibility: Oyster mushrooms are considered edible when correctly cooked.

Oyster mushroom,
Pleurotus ostreatus

In our house, we love to dip the leaves of fresh oyster mushrooms into beaten egg and seasoned bread crumbs and fry them in olive oil, or serve them in a salad with Parmigiano Reggiano, or simply offer them hot from the pan on a napkin, imagining we are enjoying cocktail hour in a Venetian bar. If you are collecting this species for the pot from the woods, study some of its extremely similar edible cousins and be sure you have a safe variety for consumption. Look for oysters that are fresh, young and firm, as they can stick around for a long time on a log even into the winter, becoming soggy, bruised, and writhing with maggots and beetles. At this stage, obviously, they are no longer appetizing. It is very important not to confuse them with their poisonous look-alikes, including the whiter, more delicate and thin-fleshed **angel wings** (*Pleurocybella porrigens*, page 151). These pretty but inedible fungi also fruit on decaying wood in shelving groups and have been reported to cause mild to serious illness. In Japan, they have even caused death in several people with underlying kidney disorders.

Pink oyster, *Pleurotus djamor,* growing from a commercial grow kit

Angel wings,
Pleurocybella porrigens

Parasol Mushroom

Macrolepiota procera

ID Features: A mature **parasol mushroom** can be over 1 foot tall, with a cap the size of a Frisbee. The young caps can be bulbous and club-like, closed tightly and attached to a membrane that protects developing gills. The mature caps have white flesh covered by flaky shingles that look brown at the tips. A prominent umbo (peak), which looks explicitly like a soft brown nipple, is visible at the top of a hemispherical cap. Under the cap, you will find white, closely spaced gills that drop white spores. Distinctively, this species has brown scale-like markings on the stipe, which develop when the mushroom shoots up out of the earth so fast it leaves stretch marks. Another distinctive feature of this mushroom is that it has a strong ring around the stem that can actually be loosened cleanly enough to move it up and down the stalk. There may be some swelling of the stipe at the base, along with pale threads of mycelium visible where it grows from the ground. However, there will not be a membranous sac around it (which is instead a sign of an *Amanita*).

Notes: I saw my first parasol mushroom in a parking lot, in the uplifted hand of a beaming amateur mycologist. It was early September, at the very first formal mushroom foray I ever attended, with the Mycological Society of Toronto. I was there to learn more about mushrooms and knew that people liked to hunt chanterelles and such for eating, but didn't take the idea of identifying mushrooms as a hobby too seriously. Upon arrival at the meeting point, I could see I was in the right place by all the large-handled baskets on top of the vehicles. I was startled by this group of Tilley-hat-and-vest-wearing club members, who seemed to be arguing with one another in Latin. So before I knew a thing about mushrooms, or became gripped by examining and arguing about them myself (sometimes in Latin!), I took photographs of eccentric mushroomers. These people, mostly older, from various immigrant backgrounds and walks of life, struck me as belonging to an odd category of unconventional human being: those who gathered urgently, sometimes with long

Parasol mushroom,
Macrolepiota procera

telephoto lenses, around small semi-rotted blobs and even turds—with fungus growing out of them. My first shot, still prominent in my files, is an image of this guy holding up a very large and handsome parasol with the pride of a first-place trophy winner.

Edibility: *Macrolepiota procera* is considered edible when correctly cooked. In North America and Europe, these substantial, sometimes abundant mushrooms are considered good (or even great) for eating. In Hungary they are especially popular and are called *nagy őzláb*, which translates to "great deer leg," making the fungus attractive to hunters of all kinds.

Unfortunately, they are not a great starting mushroom for beginner or even moderately experienced foragers because they have several very close look-alikes, including the shaggy parasol (*Chlorophyllum rhacodes*), which has a smoother-looking stem and is known to cause digestive ailments. The much more seriously concerning look-alike is the green-spored *Chlorophyllum molybdites*, which is a frequent cause of poisonings in people looking for parasols. These mushrooms, which are sometimes known as the false parasol or the vomiter, are uncomfortably similar to parasols, unless you do a spore print from the cap to see the distinctively green spores. If you want to taste parasol mushrooms, consult an experienced mushroomer in your area who is familiar with eating them.

While I think it's tricky to forage for this mushroom, the nineteenth-century British mycologist Mordecai Cooke seemed to think a person of "only dense stupidity could confound it with any suspicious species" and published numerous recipes for the meaty parasol, of which he was very fond. He suggested baking, broiling, stewing, scalloping, and potting, as well as dicing the fungus into an omelette, a ketchup, or a "Procerus Pie." Modern cooks tend to discard the tough stems of the parasol mushroom and batter and fry or stuff and grill the fresh, enormous caps.

Parrot Waxy Cap

Hygrocybe psittacina

Parrot waxy cap,
Hygrocybe psittacina

ID Features: These colorful fungi are part of a larger group known as the Hygrophoraceae, or waxy caps. The **parrot waxy cap** is quite small, with translucent emerald-green flesh. The caps are umbrella-shaped when open and have a bump called an umbo protruding from the top. They have large, widely spaced green or yellowish gills and a greenish-yellow stipe. The whole mushroom looks as if it were made of colorful wax. They are slimy when wet.

Notes: I enjoy saying the dainty word *"psittacina"* as much as I love finding these glossy jewels and their surprising hygrophorous (or watery) cousins in all colors—rose, scarlet, chrome yellow, emerald, silver, amethyst, and obsidian. Laying out your waxy caps in a group makes an impressive composition, as vivid as any wildflowers on the foray table. A foray spread without them consists mostly of shades of beige and brown.

Edibility: *Hygrocybe psittacina* are considered edible, though because they are small, tasteless, and slimy, they are not recommended for eating. I consume them greedily for their visual beauty only.

Pear-Shaped Puffball

Lycoperdon pyriforme

Pear-shaped puffball,
Lycoperdon pyriforme

ID Features: Small puffballs like the **pear-shaped puffball** can be found in large numbers growing together on rotting logs. These mushrooms are small and globe-shaped with no visible gills and have a short, squat stipe. The surface of the puff may be smooth to faintly spiky or crackled when dry. They sometimes have a visible nipple, formally called an umbo. When they are mature, they become very thin-skinned, split open at the tip of the umbo, and release olive-greenish spores like puffs of smoke. It is for this reason that puffballs are especially beloved by children with sticks.

Notes: The common names of mushrooms always have so much to say about the species' forms, colors, and other attributes, including their food value or the dangers of consuming them. In the case of the pear-shaped puffball, the common name gives everything you may need to recognize the form of them, but it's the Latin name that is so surprising. *Lyco-* means "wolf," and *pyriforme* means "pear-shaped." *Perdon* apparently means "flatulence," meaning the Latin name is essentially "Pear-Shaped Thing That Farts Like a Wolf." It's not clear who first made this association to distinguish this bulbous little ascomycete (spore-shooting mushroom), but it does give it some extra dimension and mystery.

Edibility: *Lycoperdon pyriforme* is considered edible when properly cooked. I have to admit I dislike the flavor of puffballs, but this one has been noted far and wide to be bland at best and bitter at worst. If you can't resist giving these a try and have been careful to identify them correctly, be sure to cook only puffballs that are firm and pure white throughout—even a hint of yellow or greenish spore development in the center will make them inedible. It is not true that all puffballs are edible, either; there are many toxic puffs, including the **poison pigskin puffball** (*Scleroderma citrinum*, page 159) and its cousins, which are common. The poison pigskin puffball, sometimes known as an earthball, is distinguished by its scaly, firm skin and, most dramatically, its mature purple to black spore mass inside. Slicing one open is an impressive party trick, as everyone expects to see something white, rather than this menacing and unexpected black hole.

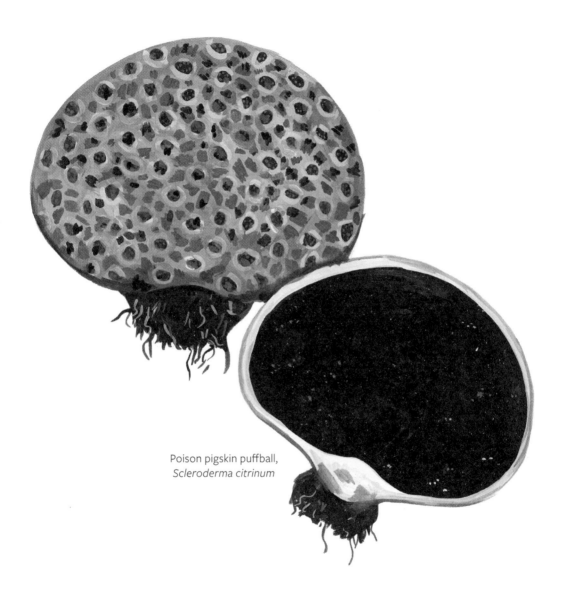

Poison pigskin puffball,
Scleroderma citrinum

Pheasant's Back

Polyporus squamosus

Pheasant's back,
Polyporus squamosus

ID Features: These bracket mushrooms are known as polypores, as they belong to a group of mushrooms that have pores instead of gills on the underside where spores are produced and dropped. A **pheasant's back** has a white, spongy underside with fairly large visible pores. The stipe is short and black at the base. The cap has brown fibrous scales that radiate outward on the asymmetrical beige bracket. These mushrooms are found on living or dead trees and stumps, fresh in the spring, and getting tougher and larger throughout the season until the end of fall.

Notes: Pheasant's back is also sometimes called dryad's saddle—as in a seat for a tree-spirit. It is a very common fungus and is widespread across North American woods. And since it does indeed look like a feathered pheasant or grouse, I have heard of hunters stalking and even shooting at the mushroom—causing much disappointment to them, I'm sure.

Among other polypore mushrooms named for birds—and considerably more exciting to find—is **hen of the woods** (*Grifola frondosa*, page 162), also known on restaurant menus as maitake. I personally find these fungi uncomfortably soapy-tasting, though they are coveted by many as both food and medicine. They are popular with photographers when they are found growing as large as a feathered bush. I know of mushroomers, particularly from the New York Mycological Society, who discreetly visit local cemeteries planted with lots of elms and oak trees (two of their mycorrhizal partners) especially to find them.

Yet another example is **chicken of the woods** (*Laetiporus sulphureus*, page 163), with its thick, tender brackets the color of egg yolks and lemons. It is a joy to find a bright young specimen bulging out of a tree like leaking sunshine. The tips of young fleshy chickens are buttery and mild when sliced and sautéed with cream,

and they look pretty on toast. Since there is nothing else that looks very similar to this polypore, it is considered one of the friendly or "foolproof" mushrooms for beginner foragers for edibles.

While sometimes tricky to identify the exact species, **turkey tail** (*Trametes versicolor*, page 163) is a common bracket with vivid, contrasting bands of color reminiscent of the tail feathers of woodland birds. While it is also probably shot at by hunters, it is mostly collected for use as a medicinal supplement in capsules and tinctures.

Edibility: *Polyporus squamosus* is considered edible when cooked. This species is easy to find, and it can be so large and striking that it is a joyous discovery for a new mushroomer. These mushrooms are only harvested for eating in the springtime, however, when very young and tender. I have eaten them a few times, but only when I had been looking for morels and couldn't find a single edible thing except for these conspicuous brackets. Even when young, however, the taste and texture of them reminded me of a tough old boot made of synthetic materials.

Hen of the woods,
Grifola frondosa

Chicken of the woods,
Laetiporus sulphureus

Turkey tail,
Trametes versicolor

Plant Pot Dapperling

Leucocoprinus birnbaumii

ID Features: A **plant pot dapperling**, sometimes known as a yellow houseplant mushroom, is a small lemon-yellow fungus you might find growing in your tropical plants. It has fine features, including a sprinkling of tiny scales on its lined, pointed cap and on its yellow stem. It also has a delicate, evanescent ring on a stem that swells gracefully at its base.

Notes: More than a few times, friends and students have sent me a photograph of a small yellow fungus growing from a houseplant in their house or apartment. They are curious about whether the mushrooms are dangerous, or psychedelic, or if there is something seriously wrong with their plants.

Plant pot dapperlings, and a few other very similar species, are not native to North America and only grow in the wild in the southern hemisphere. Spores and mycelium from these species tag along when tropical plants are cultivated, and this means they flourish in greenhouse conditions where our favorite houseplants are grown for market.

I haven't had one grow in my own plants yet, though I've been hoping to host one eventually. I consider these bright yellow fungi to be exquisite in all their delicacy and miniature detail, and revere the dapperlings (especially a few close relatives of this species we do have in the wild, including the sharp-scaled parasol, or *Lepiota acutesquamosa*) for their powerful combination of diminutive size, aesthetic perfection, a dandy common name—and sometimes an impressively powerful poison.

Edibility: *Leucocoprinus birnbaumii* is considered toxic and should never be eaten. So while these mushrooms are not dangerous to handle, I advise caution if you have children or pets (or fearless teenagers), to be sure your plant pot dapperlings are not tasted or ingested. If this isn't a risk, the fungi themselves are not dangerous to your plants (they feed on dead material in the soil) and are not dangerous to your home environment.

Keep in mind, however, if you let them mature, the spores may germinate elsewhere in other plant pots, and they can certainly fruit again. So if you really can't take the risk of having these pretty little yellow mushrooms in your plants, you might try to discard them when they first appear and even scrape out a few centimeters of soil where the mycelium lives and replace it—though this isn't a guarantee that the dapperlings won't find a way to return. If you are seriously concerned, the most reliable course of action may be to throw away (or give away) the entire plant.

Otherwise, enjoy the short visit from these small and pretty fungi, and be prompted to remember how our movement of plants and other creatures around the earth (and even into our homes) may also bring unintentional species that go on to interact with native species— and with us—in complex and unexpected ways.

Plant pot dapperling,
Leucocoprinus birnbaumii

Plums and Custard

Tricholomopsis rutilans

ID Features: I adore this uncommon and very pretty mushroom. It has a fibrous fuchsia cap with the fibrous scales along the stipe in the same shade of dark purply pink. Its flesh and gills are lemon yellow and shine through wherever the mushroom may have been nibbled or damaged in the field. **Plums and custard** grow from rotting stumps in otherwise dull conditions and radiate chroma like a sunrise.

Notes: Plums and custard is an edible-sounding and edible-looking fungus that is sadly considered to be poor quality, bitter, and even toxic for some. Keep an eye out for this ironically named wood-decaying species, though, if only for the satisfaction of pronouncing out loud the name of a cozy English pudding.

Edibility: *Tricholomopsis rutilans* is not considered edible and has caused gastrointestinal reactions in some. If you like mushrooms that are named for nostalgic baked goods—even if actually moderately toxic—look for another

Poison pie,
Hebeloma crustuliniforme

specimen, one that could have been written up as a murder weapon in an Agatha Christie novel: the unassuming, forgettably colored *Hebeloma crustuliniforme*, also known as **poison pie**.

Plums and custard,
Tricholomopsis rutilans

Raccoon Skull

When you are searching for mushrooms and looking intently at the forest floor, you will see many remarkable things. I spotted this skull when I was looking for morels in the spring, in a scrappy patch of woods near a road. It gleamed brightly at me from the end of an entire orderly skeleton lying in the dirt. The creature it came from must have died more than a year before, undisturbed but for the insects and fungi that did their quiet work of decomposition among the leaves. It is spooky to find a skull—a trope of every vanitas—with all of its portents of death. Even clean bones, free from any trace of fleshy rot, can still summon a queasiness in the gut. Despite our aversion to the fact of it, a skeleton declares that everything will die and is dying around us; the things we collect and examine are just here for a brief moment, as are we before we metamorphose, too. What violence might have come to this warm-blooded creature? Was it violence or just the endurance project, the one we share with all animals, of living itself?

Raccoons are uncannily relatable, with a spine, rib cage, joints, and digits that are only modestly different from our own. If this wild animal were alive, it would be so alert to my presence, and I would be lucky to get even a glimpse of it. I am lucky to find a skull like this one, to be close enough to note the large sockets for eyes that could see in the dark and to examine its canines and incisors, suitable for a diet as omnivorous as my own. Its curved jaw even lifts at the edges, locking it in what looks like a smile.

David Foster Wallace famously described the skull as home to a tiny kingdom. With a skull in my hands, I can't help but imagine the whole wide scope of sensing and feeling that was contained by it. We are each host to our own kingdom, within larger and larger kingdoms, with the good fortune to be alive while we are, and able to pay attention to it.

Raccoon, *Procyon lotor*

Satan's Bolete

Rubroboletus satanas

ID Features: *Rubroboletus satanas* are notoriously big, imposing fungi, with bone-colored caps that can grow to be as large as a serving platter. When clean and young, they have yellow-orange pores that turn bright red upon maturity. There is a faintly visible yellowish reticulation (a fine netting) on the bulbous yellow-and-red stipe. You might even describe a young specimen as an adorable, brightly colored orb, except shortly, a baby **Satan's bolete** begins to summon its namesake by swelling gigantically and turning bruised, bloody, and bluish, writhing with larvae and emanating its mature smell, described as "putrid."

Notes: While researching Satan's bolete, I came across a graphic time-lapse video close-up of a *Rubroboletus satanas* actively turning blue and vibrating with maggots. I could only watch the zombie grotesquerie for a moment, though I did register some delight in the ways fungi can speak to me right where I keep my mortal fears. And I delight, too, in the weirdness of a subculture that thrives on these innocent dramas, as ordinary as any other natural thing melting into compost.

Among our inheritances in the US and Canada from the UK is the fear of mushrooms—sometimes called mycophobia—stemming in part from the mysteries of mushroom growth and reproduction in the dark, but no doubt acquired while watching the mushroom-eating victims of abject poisonings, wild hallucinations, and painful deaths. I am intrigued by the fact of so many mushrooms named for demons, like this one. Others include a black cup fungus called the **devil's urn** (*Urnula craterium*, page 173), a tentacled stinkhorn that looks like an octopus charred by hellfire itself, called **devil's fingers** (*Clathrus archeri*, page 173), and the outrageously pink, shiny, and phallic *Mutinus caninus*, sometimes known as the dog stinkhorn but also called (first by our modest Victorian ancestors, surely) the **devil's dipstick** (*Mutinus caninus*, page 172).

Edibility: *Rubroboletus satanas* are very poisonous and are not edible. While there are many species of boletes that are collected for eating, everyone knows to avoid any boletes that have red pores or turn blue upon bruising. This striking

Satan's bolete,
Rubroboletus satanas

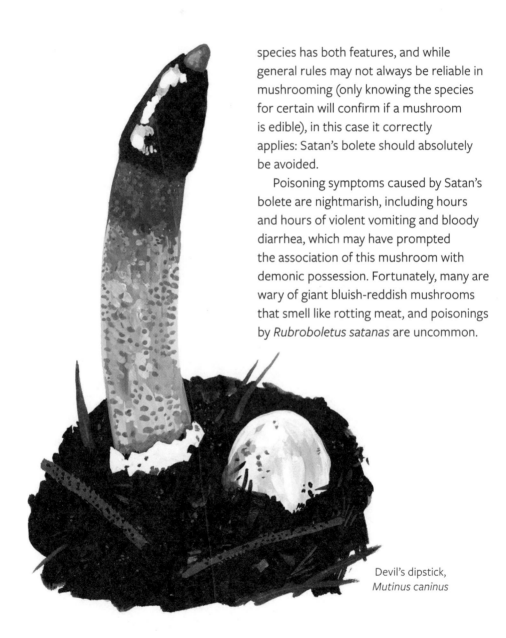

species has both features, and while general rules may not always be reliable in mushrooming (only knowing the species for certain will confirm if a mushroom is edible), in this case it correctly applies: Satan's bolete should absolutely be avoided.

Poisoning symptoms caused by Satan's bolete are nightmarish, including hours and hours of violent vomiting and bloody diarrhea, which may have prompted the association of this mushroom with demonic possession. Fortunately, many are wary of giant bluish-reddish mushrooms that smell like rotting meat, and poisonings by *Rubroboletus satanas* are uncommon.

Devil's dipstick,
Mutinus caninus

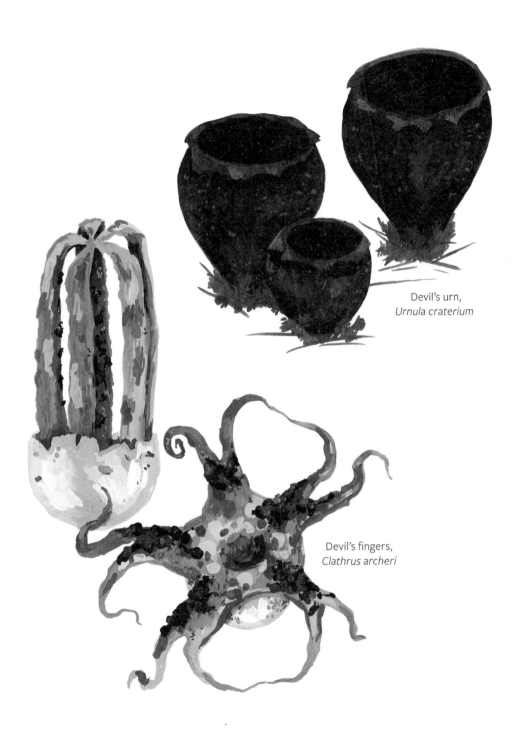

Devil's urn,
Urnula craterium

Devil's fingers,
Clathrus archeri

Scaly Pholiota

Pholiota squarrosoides

ID Features: The **scaly pholiota** is an easy mushroom to remember, as it grows from living and dead trees in striking bouquets. The fungi are pale-fleshed, with a membranous ring that may or may not be present on fully opened specimens. They have rusty-brown spores and stain brownish when bruised or aged. Most distinctively, these pholiota are covered with sharp, rusty-brown scales all over the slimy cap and the stipe.

Notes: Wandering off forest trails, hyper-focused straight at the ground, happily following a trail of *possible* mushrooms, is an activity that seems intentionally designed to make someone lost. As the kind of person who can get lost anywhere, I have learned to always be alert to landmarks. While in the woods, I recommend the sturdy *Pholiota squarrosoides* as a good candidate with which to navigate by mushroom. If you see a bunch, make a note of the location. And because they are so graphic and camera-friendly, groups of students also like to take photographs of themselves alongside them, with the students—just the like mushrooms—arranged tightly in a cluster.

Edibility: *Pholiota squarrosoides* is not recommended for eating. While the scaly pholiota, if correctly identified, may itself be safe to consume, it looks much too much like its poisonous cousin, the shaggy pholiota (*Pholiota squarrosa*). Not only is it tricky, no redeeming flavors have been reported for this mushroom to make it worth the risk.

Scaly pholiota,
Pholiota squarrosoides

Scarlet Cup

Sarcoscypha austriaca

Scarlet cup,
Sarcoscypha austriaca

ID Features: The **scarlet cup** fruits on bits of dead wood, usually on the ground under leaves in the spring. The inside of the cup is a bright, saturated red, while the outside of the smooth cup is paler and attached to a very short stalk. Even at close range, the thin cup will look much like a velvety rose petal.

Notes: In the early spring when you hope to see the first morel, the trees are bare and black and the ground is covered in the rotted vegetation of last fall. There are very few other interesting or colorful fungi at this time, or any signs of new life at all. In these moments, especially if you haven't found any prized edibles, it is startling—and a consolation—to encounter a

Orange peel fungus,
Aleuria aurantia

gleaming red *Sarcoscypha austriaca* in the muck.

If, instead, you find a bright orange cup growing in wavy folds, you might be looking at the aptly named **orange peel fungus**, *Aleuria aurantia*. More than once, I have excitedly spotted one only to discover that it was an actual orange peel, from a discarded lunch.

Edibility: As tempting as they might seem, scarlet cups are not considered edible.

Since this species can look a little like a glass of wine on its short stem, I do enjoy thinking of a widely scattered swath of them as the remains of Bacchanalian elvish parties.

Shaggy Mane

Coprinus comatus

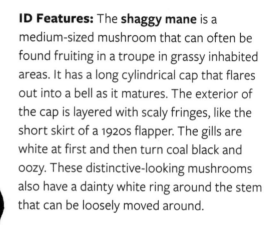

ID Features: The **shaggy mane** is a medium-sized mushroom that can often be found fruiting in a troupe in grassy inhabited areas. It has a long cylindrical cap that flares out into a bell as it matures. The exterior of the cap is layered with scaly fringes, like the short skirt of a 1920s flapper. The gills are white at first and then turn coal black and oozy. These distinctive-looking mushrooms also have a dainty white ring around the stem that can be loosely moved around.

Notes: Shaggy manes belong to a family of *Coprinus* species known as inky caps, because the caps dissolve quickly into a black, spore-filled liquid. While the natural process has a graceful name that I relish saying aloud—*deliquescence*—the ink itself is made by the unappealing process of trapping your handful of darkening mushrooms in a jar for a day or two, until the decomposing mass can be strained, stabilized and deodorized of its putridity.

Edibility: *Coprinus comatus* is considered edible when correctly cooked. While it is often recommended as an easy mushroom for beginning foragers, there are numerous reports of people confusing it with poisonous

Shaggy mane,
Coprinus comatus

species of *Amanita* or *Chlorophyllum*, or with *Coprinopsis atramentaria*—a black-spored, deliquescing cousin of the shaggy mane. *Coprinopsis atramentaria* is also known as **tippler's bane**, since it is safe to eat, unless you drink alcohol up to three days after eating it! This mushroom contains a chemical similar to a pharmaceutical used to treat alcoholics, causing unpleasant flushing and other symptoms. For these reasons, to be certain, beginners should study, collect, and investigate shaggy mane mushrooms several times before consuming them.

Even though I have been mushrooming enthusiastically for many years and feel confident I can identify one, I admit I have not yet tasted a shaggy mane. I have collected them, however, and even the young specimens deliquesce all over my hands and the basket, turning everything unappetizingly stained and grey by the time I get them home to the kitchen.

Coprinus species are made diminutive and endearing by their nicknames like "shaggy" or "inky," but they are ghoulish shapeshifters in their own right. Also, since they tend to grow in disturbed ground, along roadsides and in backyards, they can

Tippler's bane,
Coprinopsis atramentaria

accumulate toxins from the environment and become contaminated. Despite all of this, they are familiar edibles to many, considered mild-tasting and earthy (like brown cremini mushrooms at the supermarket), and are often used in soup. I have also read of cooks deliberately squeezing out the black liquid to use like squid ink, to stain dough for a dramatic mushroom pasta.

The Sickener

Russula emetica

ID Features: The sickener is one of many *Russula* species, which are also known as the brittlegills. Mushrooms in this genus do indeed have brittle gills that crumble when you rub them into shards that resemble flaked almonds. The cap is usually convex and can sometimes have a faint bump (or umbo) in the middle. A young sickener cap can be a vivid five-alarm red and may be sticky with slime when wet. The stipe and gills are pure white, with areas that are sometimes faintly pink.

Notes: Red brittlegills are notoriously hard to identify by appearance alone, and senior mushroomers will sometimes nibble one or ask a bold (or naive!) new mushroomer to confirm whether the specimen tastes hot and acrid and produces a tingle on the tongue. If it does, you may have a sickener, as opposed to any of a number of other beautiful red *Russula* species with names like scarlet, rosy, gilded, or bloody brittlegills.

Edibility: *Russula emetica* is considered poisonous. The sickener is a mushroom that lives up to its name, in any language. *Emetica* means "causing vomiting" in Latin, so this species should be avoided. Specimens have been known to cause oral discomfort and gastrointestinal illness. Even though some other red *Russula* species are considered edible, and they might look as cheerful and harmless as polka dots, foragers are strongly discouraged from eating any red brittlegills.

The sickener,
Russula emetica

Skirted Stinkhorn

Phallus duplicatus and *P. indusiatus*

ID Features: The **skirted stinkhorn** and its cousins, including the pointed, pink dog stinkhorn, are embarrassingly phallic in their form. Not only do they look long and turgid in the stipe, with a glossy perforated cap, they grow from testicle-sized stinkhorn "eggs" in a rapid and dramatic erection.

The skirted varieties have a miniskirt to bridal gown–length netting that grows out from below the cap, depending on the species. The **tropical skirted stinkhorn**, *Phallus insidiatus* (the one everyone knows from viral time-lapse videos on the internet), is pictured opposite with a long skirt; the North American *Phallus duplicatus* has a shorter, less dramatic netting around the stipe. The most dramatic identification feature, however, is the overpowering smell of this fungus when fresh and coated with spore-rich slime: like a horrible sewer, backed up with excrement and rotting meat.

Notes: Stinkhorns are unwelcome in anyone's yard. I have heard reports of plumbers and natural gas inspectors being called in to identify the source of a stench, only to find one of these fungi exhibiting itself gruesomely in the grass. The iconic American mycology professor Tom Volk says he gets more inquiries about how to get rid of stinkhorns than almost any other mushroom question, and his best advice is to pave the yard over! When I lead mushroom forays for students, they frequently find stinkhorns and bring them back proudly in their *bare hands* to put them on the foray table or on the roof of my car. Stinkhorns disperse their spores by attracting flies to their smelly caps, and together with my students, I have seen flies gorge on all of the spore-slime from a stinkhorn, leaving it clean and pale in a matter of minutes—thankfully! And yet, if I'm within 150 feet of one, I cannot get the smell out of my sinuses, my lungs, or the darkest death-fearing depths of my imagination for several hours.

Edibility: According to many of my field guides, stinkhorns are widely considered inedible, and at the very least, extremely unappetizing. Historically, corpse-smelling stinkhorns have been associated with witchcraft and superstition, but because of their penis-like appearance, they have also been ingested as aphrodisiacs. Charles Darwin's daughter Etty is said to have hunted stinkhorns with a pointed stick and

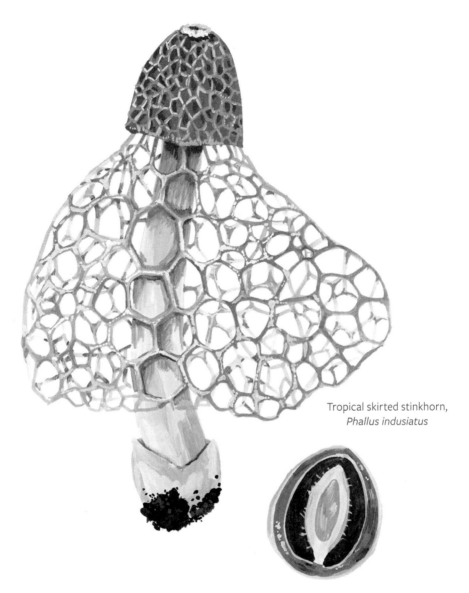

Tropical skirted stinkhorn,
Phallus indusiatus

burned them in a locked room to protect "the morals of the maids."

There are some stinkhorn species that are eaten in Europe and Asia today, and I know fellow mushroomers in Toronto and New York who have tried pickled stinkhorn eggs, which they said were not stinking at all, but crisp and spicy-tasting, like a radish. Washed and dried stinkhorns (which look and feel like loofahs) can be found for sale in Asian markets around the world.

Sleeping Beauty
(Chrome-Footed Bolete)

Harrya chromapes

ID Features: *Harrya chromapes* is in the boletus family, as it has white spongy pores under its cap instead of gills. It has a dry dusty-rose cap and small dark-pink scales dotting the top half of a white stipe. The lower half of the stipe surface is bright and starkly yellow.

Notes: I've always called this mushroom *Tylopilus chromapes*, though before it got this name, it was known as *Boletus chromapes*, and before that, *Leccinum chromapes*. I have since learned its name was recently changed again, this time to *Harrya chromapes*, in order to group it together with other species that have molecular features in common.

Modern DNA analysis has changed mushroom naming in ways that are maddening for older mycologists who learned their fungi by their visible features and memorized the old Latin names. Michael Kuo has a rant about the subject for boletologists in particular, calling these taxonomic contortions a "ridiculous display of career making."

The only part of the name in common is the *chromapes* (*chrom*: chrome yellow; *pes*: foot), a yellow-foot mushroom, which makes at least part of the name easy to remember. Commonly this pink-and-yellow bolete is known as the **chrome-footed** or **yellow-footed bolete**, which is descriptive but not very evocative of the surprise this pretty mushroom will produce when found in the dark duff. Hence the new common name I am proposing here, based on a suggestion by my friend Esther, who is seven and dreams of finding her own **sleeping beauty** in the forest. You will be as delighted as Esther when you find one, and I won't tell anyone if you must kiss its velvety pink cap.

Edibility: *Harrya chromapes* is considered edible when cooked. It is a good, if mild-tasting, edible mushroom, though look for specimens not totally riddled with fungus gnats and fly larvae. Like many boletes, its flavor can be concentrated by drying it first and rehydrating it for cooking. Beware of poisonous boletes and many similar look-alikes that may not be edible, or whose edibility may not be known.

Sleeping beauty
(chrome-footed bolete),
Harrya chromapes

Finnish forest reindeer, the species
used in the exhibition *Soma*

SOMA

Installation art by Carsten Höller

HISTORICALLY, INDIGENOUS PEOPLE in Finland and Siberia have eaten the fly agaric—that iconic red-and-white mushroom—for recreational and spiritual purposes. Though the veracity of these stories is disputed, it is said that people would eat the meat of reindeer who were eating the hallucinogenic fungi in the wild, and even collect and drink their *urine* in order to feel the mental effects of the mushroom, while avoiding some of the more unpleasant gastrointestinal symptoms brought on by direct ingestion.

The contemporary German artist Carsten Höller explored this sensational history in an installation at the Hamburger Bahnhof museum in Berlin entitled *Soma*, in which reindeer, mice, canaries, and flies occupied a large gallery with these mushrooms. The public could watch the behavior of animals affected by the toxic fungi and see samples of their muscarine-infused urine stored in small fridges. Some gallery visitors were eligible to win an opportunity to spend the night in the exhibition, when the fridges were unlocked and they were offered the reindeer urine to drink, allowing guests to experience the effects of the diluted toxins on themselves. While it wasn't clear which bottles of urine were from mushroom-eating reindeer or were other kinds of liquid, gallery staff claim that more than a few people did indeed "drink the mini-bar dry."

Carsten Höller describes the experience of ingesting the muscarine poison as largely unpleasant: "The first four times I tried it, I became comatose. Then you wake up, throw up, and you don't know where you are . . . the sixth time I started to chant like a Tibetan monk."

Verdigris Agaric

Stropharia aeruginosa

Verdigris agaric,
Stropharia aeruginosa

ID Features: It is incredibly rare to catch this mushroom in its most vivid swimming pool–blue stage. This *Stropharia* is uncommon and only appears in this color while the mushroom is very young. At this stage, it is slimy and woolly all at once, with a fibrous ring on the stipe. The fungus quickly fades to a greyish-yellowish and shows a more developed bump (or umbo) on the *cap*. The gills of a **verdigris agaric** are purple grey and darken to purple black.

Notes: The name of this mushroom refers to verdigris, the crusty bloom on oxidizing copper or brass that shares the same unmistakeable turquoise color. My friend, the artist Hannah Rowan, who makes elegant gallery installations of ceramic vessels, pipes, flowing water, and accumulating minerals, calls the verdigris in her glazes "the color of impermanence."

Edibility: *Stropharia aeruginosa* is not considered edible. However, the **wine cap** (*Stropharia rugosoannulata*, page 189) is a much more common cousin of the verdigris, and you might find this sizable mushroom growing in your garden mulch.

Wine cap,
Stropharia rugosoannulata

It has a distinctive tawny-burgundy color on the cap, a spiky collar-like ring on the stipe, and purple-black spores when mature. Unlike the verdigris agaric, wine caps are a popular wild edible when you know how to identify them correctly. Or grow your own from a kit, as they are available commercially.

White Elfin Saddle

Helvella crispa

ID Features: A *Helvella crispa* is a graceful specimen. The cap is made of pale thin flesh that flares upward in wavy folds—less like a saddle, perhaps, than a blown-out skirt. It has no gills, as it's a sac fungus, closely related to morels and cup fungi. It has a distinctive stipe made of numerous ridged folds that look to me like a bundle of pale, rinsed tendons. A cross-section of the stipe will also show many irregular hollow chambers. **White elfin saddles** are mostly found in the spring, growing on the ground in the woods.

Notes: The white elfin saddle is one of many mushrooms associated with elves and other sprites, including fairies, pixies, witches, angels, and demons. Emily Dickinson wrote a poem known as "The Mushroom Is the Elf of Plants" about the way mushrooms grow speedily in the dark and surreptitiously emerge in places that were without a trace of them a moment ago. You can imagine that before it was known how fungi could obtain nutrients without the sun and reproduce without seeds or roots, people would turn to the supernatural for explanation. In the poem, Dickinson accuses the fungus of being nature's ultimate betrayer and a mischievous imp, too, because in the end the mushroom devours all.

Edibility: *Helvella crispa* is not considered edible and may be poisonous. Foragers might be more familiar with the darker versions of the elfin saddle, including the more common *Helvella lacunosa*, which is sometimes eaten with caution, after cooking well to remove any volatile toxins found in similar mushrooms.

White elfin saddle,
Helvella crispa

Witch's Butter

Dacrymyces chrysospermus

Black witch's butter, *Exidia glandulosa*

Description: Jelly fungi that are commonly called **witch's butter** might be yellow, yellow orange, orange, or black. *Dacromyces chrisospermus* is a yellow-orange version that is sometimes also called orange tree brain. It emerges crumpled in bunches and rows on living or dead trees, with a gummy-bear texture and translucency. The jelly fungus is also around gummy-bear size, though irregularly contorted and wavy.

Notes: I thought I knew what witch's butter was, and have been pointing it out to students and beginners for many years on forays. It's an easy mushroom to spot, especially when yellow, and quite common. Just as I was preparing to write this section, I saw a vast range of species belonging to numerous genera—*Tremella lutescens, Tremella mesenterica, Phaetotremella frondosa, Dacrymyces palmatus* and *Exidia glandulosa* (for the black variety) among them—that are all known in some text or other as witch's butter. Many are hard to distinguish in the field for anyone other than keen survivalists or jelly fungi aficionados. I have always held some affection for this range of mushrooms, and I like to tell children

that it is the stuff witches spread on their toast. An unaccounted-for legend has it that if this mushroom is growing from your door frame—or as it is for us right now, from the wood chopping block outside our front door—it may signify a curse on your home! But all the witches I know have been the most uncommonly good people—they revere animals and flowers and parties—so instead of falling for the old stereotypes, I take its presence at my doorway as a gift, a benevolent enchantment.

Edibility: *Dacrymyces chyrsospermus* is considered inedible. While some field guides describe the range of witch's butters as inedible, some are described as edible, but insubstantial, watery, and tasteless. I have seen hikers nibble on a bit of *Dacrymyces chrysospermus,* and some do consider it to be an emergency survival food, but it hasn't been studied or accounted for in this way, and I could not recommend it.

Witch's butter,
Dacrymyces chrysospermus

Witch's Hat

Hygrocybe conica

ID Features: When mature, a **witch's hat** mushroom is pointed like its namesake and even flares out jauntily at the base of the cap. It is another of the small waxy cap mushrooms, with flesh that looks like brightly colored, translucent wax and is slimy when wet. The cap is yellowish orange to bright poppy red, with pale whitish-yellow gills and white spores. The stipe is yellow. The entire mushroom is distinct from other red waxy caps because it turns inky black with age or when bruised—and possibly, when kissed by witches.

Notes: Kim Kozzi and Dai Skuse are the much-beloved duo known in contemporary art circles as FASTWÜRMS, an ongoing collaborative project grounded in do-it-yourself methodologies and Witch Nation identity politics, in solidarity with working class, queer, and other marginalized creative communities. I asked my friends what the pointy witch's hat means to them, as they are often seen wearing them in the public performances and videos they make. Kim told me, "There are a lot of different reasons for the hat to be linked to Witches"—as they refer in part to the tall pointed black felt hats found on mummified women from the Iron Age in Subeshi along the Silk Road in China—"but for us, it is about a continuation of very old traditions linked to ceremony and social display." Kozzi and Skuse describe Witchcraft as a "liberation path" and a knowledge tradition, less like a religion and more akin to Zen Buddhism. Kim also described using the iconic hats for special occasions involving directed divination, meditation, lucid dreaming, and ritual fasting.

The witch's hat is one among many fungi named for supernatural beings—like demons, angels, and fairies. **Fairy ring mushrooms** (*Marasmius oreades*, page 196) are also in this company, because they show up in groups, all fruiting from the same central mycelium, growing at the edge of uncanny, perfectly round circles in grassy fields. It's not surprising that a mycophobic (mushroom-fearing) culture emerged in North America, with settlers who were ignorant of local flora and fungi and terrified of hallucinations, illness, and deadly poisonings. And since mushrooms have the surprising ability to spring up rapidly in the dark and sometimes in weird, suggestive bodily forms, an entire kingdom was seen with the prejudices often directed at witches. Mushrooms were then, and by

Witch's hat,
Hygrocybe conica

many are still, deemed dangerous, vulgar, and uncomfortably mysterious.

One of my favorite drawings made by the FASTWÜRMS—or the 'würms, as they are affectionately called—is of silhouetted figures in pointy black hats gathered around an inhabitable fly agaric mushroom. The text below it declares PROPS TO THE FAIRY PEOPLE, calling on the world to recognize and honor the queers, the misfits, the artists, and the witches.

Edibility: *Hygrocybe conica* is not considered edible, and there has been at least one poisoning attributed to this species. And it may go without saying, but any mushroom that turns to black slime when cooked should not be eaten.

Although edible, fairy ring mushrooms can be difficult for beginners to identify with certainty, and have many poisonous look-alikes, including many other mushrooms that grow in rings. However, when it is correctly identified, *Marasmius oreades* is a popular edible mushroom.

Fairy ring mushrooms,
Marasmius oreades

Wood Ear

Auricularia americana (formerly *Auricularia auricula-judae*)

ID Features: **Wood ears** are common fungi found growing on the branches of deciduous trees, especially American elders. The tops of the stipeless caps are tan to coppery brown and covered in very fine grey hairs. The inner surface of the convoluted ear-shaped fungus is smooth rather than gilled, pale, and shiny. They are firm, rubbery, even gelatinous when wet. Sometimes specimens are also finely wrinkled. Depending on the specimen and the weather you find it in, a wood ear might be as silky and wavy as a flower petal or look like a misplaced flap of slimy, veiny flesh.

Notes: The former Latin name for this mushroom refers to the Christian apostle Judas, who is said to have hanged himself from an elder tree after betraying Jesus. Over time, the name for this gnarly arboreal excrescence was misappropriated from "Judas" to simply "Jew," and the most unflatteringly bigoted of the mushroom names was coined, one that I still hear used too often: "Jew's ear." Many people now use other names for this fungus, like wood ear or jelly ear.

Sadly, every time I hear the old common name, and others including the pejorative "gypsy" for the ragged but edible **granny's**

nightcap (*Cortinarius caperatus*, page 199), I am reminded not just of the lack of knowledge or sensitivity of the speaker but also how many shameful stories of intolerance are written right into the names of our plants, animals, and fungi.

European naturalists in the eighteenth century contributed to the racist denigration of the body parts of people in one group or another, informed by the practice of studying visible—or what is known as morphological—characteristics. The early researchers compared colors and forms of creatures to distinguish among species and to understand their connections with one another. While it was a reasonable approach at the time and resulted in many meaningful insights about the nature of relationships and evolution among species, since then microscopy and modern DNA-mapping technologies have forced a good deal of morphological order to be questioned and revised. Unfortunately, arbitrary visible traits in people—like skin color, facial features, and the size of one's head and other body parts—have been used to categorize and evaluate human beings, a concept that has been dangerously appealing to those eager to profit from treating people as inferior and exploiting them.

While all our modern nature hobbies—like mushrooming—are informed by many generations of amateurs and scientists who provided detail and insight into the operations and wonders of the natural world, it is important to be mindful of the subjectivity and power that describing a species can hold. Naming is a cultural act. It can be narrow and judgmental—or, alternatively, open-minded and empathic. And our recorded names for things give us not just an understanding of objective characteristics of individual specimens but also a record of who we are as human

Wood ear,
Auricularia americana

beings and how we live together and with other creatures in the world.

Edibility: *Auricularia americana* is considered edible when correctly cooked. It is a popular mushroom in soups and stews, and big bags of it can be purchased dried in any Chinese market. They have a firm, rubbery texture and a generous woodsy flavor (rather like black trumpets), and they get slippery when rehydrated; I would even say they are crisp when chewed. You may find them in Chinese soups, shredded on Japanese ramen, or stir-fried with delicious, stretchy Korean glass noodles—a great experience of so many textures in a dish! Medicinally, the wood ear has a fascinating and long history, with records of it being used in both the East and the West for centuries to treat everything from sore throats to jaundice and eye diseases. Modern clinical trials are showing that compounds in the fungus may be useful for pharmacological applications in the treatment of blood ailments and to lower cholesterol.

Granny's nightcap
(as it is known to the Finnish),
Cortinarius caperatus

Yellow Morel and Black Morel

Morchella esculenta and *M. conica*

ID Features: Morels look maddeningly like pinecones and expertly camouflage themselves in the leaf litter in spring among hardwoods, especially elm, ash, maple, oak, and apple trees. When I am hunting for them on the forest floor, the image in my mind of a morel is so vivid and powerful, and I have a sense that my brain sees the honeycomb-pocked cap and the cream-colored stipe before my eyes do.

At different stages of growth, **yellow morels** and **black morels** look very similar, and their caps can change in tone and color considerably, showing up in various shades of cream, butter yellow, tan, or brown. Yellow morels tend to have more rounded, lighter caps (especially lighter on the borders between the dents), and black morels are pointier and can look almost black. If you slice open a true *Morchella esculenta* or *Morchella conica,* you will see that the stipe is completely hollow, and it folds and gathers like drapery. The cap is attached to the stipe in one continuous line all around, and it is completely hollow, too. While you might find a solitary resident centipede, there are no chambers or other substantial flesh tissue inside a morel, and this is one of a few important

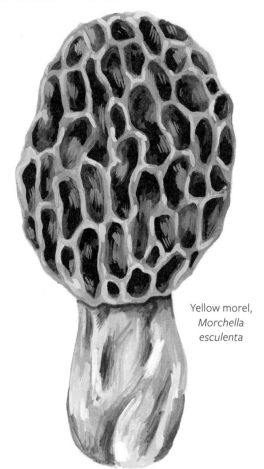

Yellow morel, *Morchella esculenta*

features distinguishing them from their less palatable—and even toxic— look-alikes.

Notes: Yellow and black morels are so beloved, known and sought after that more than a few inexperienced foragers

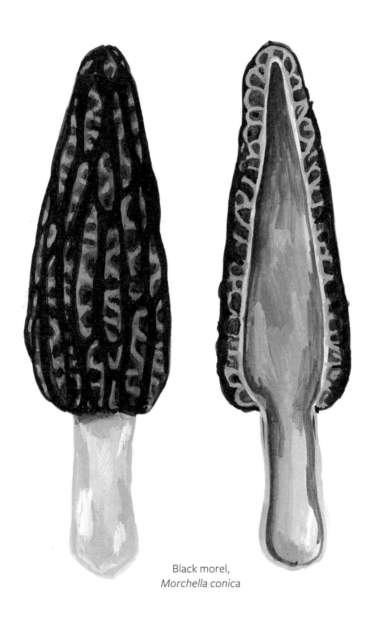

Black morel,
Morchella conica

and careless chefs have either confused them with toxic look-alikes or served their true morels undercooked or raw—a practice that can cause some serious digestive distress. Even though some of the look-alike fungi that also grow in the springtime are eaten by some people, especially in Eastern Europe, the **false morel** (*Gyromitra esculenta*, page 203) is considered toxic, and for some unfortunate diners—who may have been eating false morels happily for years—it suddenly causes poisoning and even death. Even cooking a false morel without substantial ventilation can be dangerous, as it contains a compound that can be converted into the volatile toxin monomethylhydrazine—a chemical used by the military as a rocket propellant!

Early morel is a common name for other fungi related to false morels that can also cause digestive issues and are not recommended for eating. If you examine these look-alikes closely, you will find they may have chambered interiors and a cap that is partly free of the stalk—signs they are not true morels. It can be confusing for new foragers, especially if you search the internet for advice about the edibility of *Verpa bohemica* or *Verpa conica* (sometimes called early morels, or thimble morels) because despite all the warnings and real accounts of poisonings, people sometimes eat them and swear they enjoy them anyway.

Edibility: *Morchella esculenta* and *Morchella conica* are considered edible, but only when fully cooked. As with any edible species, if you will be eating morels, be sure to learn the difference between the questionable toxic fungi described here and true morels. Consult an experienced forager and trusted guides, cook your specimens thoroughly and consume only a little of any edible mushroom if it is your first time trying it.

Though elusive where I live, yellow and black morels are gathered by the pailful in some regions, and there is a roaring commercial trade in specimens found in forests and burn-sites in the Pacific Northwest. In this case, they are often dehydrated to preserve them—though I have never enjoyed tough rehydrated morels as much as the fresh ones I find and eat right away for dinner. If I have extra fresh ones, I will sauté them lightly to reduce their moisture and put small packs of them in the freezer for later use in pastas, risotto, or pizza. Morels are so strange and precious—and smell subconsciously of the most intimate skin, to be honest—that if you want to endear yourself to someone, you can also give away your treasures. But be warned, such a gift could make them fall in love with you.

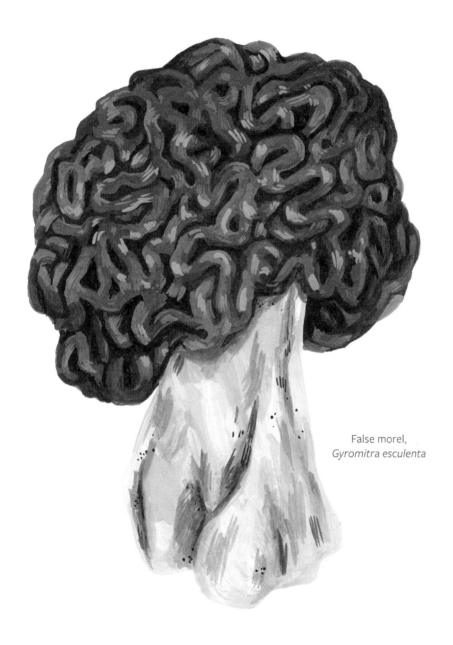

False morel,
Gyromitra esculenta

Yellow Swamp Brittlegill

Russula claroflava

ID Features: The **yellow swamp brittlegill** is one of many *Russula* species that are known for their stiff gills that crumble into shards resembling flaked almonds when you rub them. They do grow in moist, boggy, swampy woods, though they might also be found in drier areas under conifers and birch and aspen trees. The cap is usually convex, but as it matures it can sometimes be depressed in the middle. The cap is a bright, cheerful yellow and may be sticky with slime when wet. The stipe and gills are mostly white and, importantly, turn grey when sliced or handled.

Notes: When fresh, *Russula* species are tidy-looking, symmetrical mushrooms with perfectly round caps. They come in a wide range of colors, including white, bright yellow, ochre, olive, primrose, scarlet, purple, charcoal, and even black, and shine like Skittles candies from the forest floor. Other distinctive species include **green-cracking russula** (*Russula virescens*, page 206) and the slippery, peachy-colored species *Russula vesca*, which is sometimes called the **flirt** (page 207). While the yellow swamp brittlegill species is less common, brittlegills as a group are very common and widespread across North America. While I rarely eat brittlegills, I do love them for their colors and their familiar company when I'm walking in the woods.

Edibility: *Russula claroflava* is considered edible when correctly cooked. *Russula* species are notoriously hard to identify by appearance alone. While there are several yellow *Russulas* species, the edible yellow swamp brittlegill is distinctive in that the flesh turns grey when broken, and it has a mild/bland flavor if a tiny bit of a fresh specimen is nibbled and spat out. This mushroom is considered good-tasting, though field guides to edible fungi describe it as difficult to identify because there are so many mushrooms that it could be mistaken for—including toxic *Russula* species and even poisonous yellow *Amanita* species. Look for Russian or Polish recipes for this sweet, nutty-tasting specimen, as this is one of several brittlegills that are joyfully collected and consumed in Eastern Europe.

Yellow swamp brittlegill,
Russula claroflava

Green-cracking russula,
Russula virescens

The flirt,
Russula vesca

Oyster mushrooms growing from a "mycomorph" artifact
in the installation *Psychotropic House: Zooetics Pavilion of
Ballardian Technologies* by Urbonas Studio, 2016

ZOOETICS: PSYCHOTROPIC HOUSE

Installation art by Urbonas Studio

THE FIRST SERIOUS RESEARCHERS I have ever heard declare that mushrooms were from outer space—without a twitch of irony—were Lithuanian artists Nomeda and Gediminas Urbonas, in a formal lecture about their work at the Banff Center for the Arts in 2018. They said, "For us, growing up in Lithuania, the cosmology and cosmo-technics produced by wild mushrooms were part of everyday life, as was wild mushroom hunting." As for many others from the mycophilic Baltic countries, mushrooms are considered magic. Sometimes they are tricksters and other times benevolent spirits of the forest that must be respected with offerings and with rituals like turning in circles while harvesting. To the artists and others, fungi bring good luck, which holds especially true in that caring for the environments where they grow tends to ensure one's own happy future.

Much like Lauren Fournier's *Fermenting Feminism* (page 77), the Urbonas' studio work (under the larger research title "Zooetics") draws from multiple disciplines—from philosophy to speculative fiction to environmental sciences—to explore our creative relationships with non-human species. One of their many gallery installations on the theme was *Psychotropic House* (displayed at the 32nd São Paolo Biennial), inspired in part by a 1970s-era collection of science fiction stories by J. G. Ballard called *Vermilion Sands*. In the book, human-made objects are growing and living, and a sentient house can communicate and empathize with its inhabitants. In the museum, Urbonas Studio explored the potential of creating new building materials by cultivating them with living mycelium.

A mycelium is a branching network of fibers that is the real body of a fungus, from which mushrooms are the reproductive fruit. Fungal mycelia are what make up a mind-exploding system that transmits signals and exchanges nutrients between trees and mushrooms, animals, and plants, and affects literally everything dead or alive in the forest. The soil is full of it, and life as we know it cannot exist without the work of mycelium, in what is sometimes referred to as the Wood Wide

Web. Despite being as slender as spiderwebs, fungal threads are also strong, grow quickly, and bind with the substrate they are growing from. So in the art-science laboratory constructed in the *Zooetics Pavilion of Ballardian Technologies*, participants cultivated their own living materials and artifacts in collaboration with the mycelia of autochthonous oyster mushrooms. In the gallery, plastic sheets sealed in a relatively sterile environment for shelves of growing matter, which took the shape of molds full of inoculated coffee bean husks, corn pollard, and sugarcane bagasse that the artists note was gathered from plantations established by settler colonialists in São Paulo.

To commercial experimenters with this biotechnology, there is great economic interest in lab-grown mycelium-based bricks, organic packaging and resilient vegan substitutes for leather, among other applications. But for Gediminas and Nomeda, informed by love, respect, and a speculative imagination, the protypes they grow—called mycomorphs—function as proposals for radical new ways to think about ourselves and how to live in the world. Humans are—like mushrooms—inextricably enmeshed with other life forms and our environments. We are not actually superior and separate or justified in dominating and exploiting the landscape for our own individual needs. Like the Urbonases, who hold practical realities and extra-rational thinking in fluid tension, we would all be better advised to dream up new ways of living, with creativity and humility, in environments we all rely on and share.

How to Host a Mushroom Foray

Y OU CAN COLLECT MUSHROOMS any time you notice them in the grass or the woods, as long as it's not discouraged. A mushroom foray, however, is an organized mushroom-collecting event performed with others for educational or scientific purposes, and/or to collect edibles "for the pot." By harnessing the efforts of a group of people, you can gather an impressive range of species in a short time and learn a lot from this sampling about what comes up in your area and how one species compares with others. When a regional foray is repeated at different times of the year and in subsequent years, you can deepen your knowledge of a place and how it changes in different seasons and weather, and across time.

+ + +

These are the steps you will need to take to host a mushroom foray in your area.

Scope out a site: Find a forested area where you can get permission to collect specimens. As long as you have enough uncultivated woods to walk in with your group for an hour or two, you have a good site. A wet and mossy area near water is preferable to a dry one, but wherever there are trees, there will be mushrooms. Many private owners and public stewards won't mind if you collect fungi on their land, either because they don't notice or care about mushrooms or because they might be curious to see what you can find and identify in their woods. Acknowledge the privilege by sharing insights, species lists, and edibles, if you find them.

Find someone who knows more about mushrooms: Ask around to find a mushroomer with more experience and knowledge than you have to join or even lead the group. A senior member of a local mycological society is a great choice, but any modest mushroomer who admits to their limits and is cautious about confirming species for eating is a safer companion (and a more enjoyable one to talk to) than an arrogant one. See if anyone else knows about birds or geology or medicinal plants or more, and enjoy a broad range of conversations that deepen everyone's experience of a site.

Invite others to join you on the foray: Ask just one other person to keep you from getting lost in the woods, or up to 30 others, depending on the aspirations of the foray and the size of your site.

Tell everyone to pack:

- clothes for hiking in the woods in all kinds of weather, with long sleeves and pants to avoid ticks and scratches
- closed-toe shoes or boots
- a basket and paper bags for edible species (not plastic bags: they melt watery mushrooms into slime)
- a folding pocketknife (it is dangerous to carry a pointy kitchen knife)
- water
- field guides
- snacks to share

Plan to meet for about four hours: You will need about 30 minutes to gather and make announcements, one to two hours to forage, and 30 or more minutes for identification and discussion. Arrange for more informal time afterward to have a picnic and, if you are lucky, to swim.

The leader should bring extra field guides and a few extra baskets: People always show up without them. I notice that if someone is not carrying a basket, they can forget what they are doing and get distracted from the task of collecting mushrooms. A basket gives you a role to play. I always keep the cardboard fruit vessels that come with handles to bring to forays with students. And I save wicker baskets any time I see them.

The leader can also bring a blanket or tablecloth to throw on the ground and display all the collected fungi at the end of the foray, so everyone can gather and see the mushrooms. A folding table is best if you have one and are willing to carry it into the woods. I have also used the hood of my car for the ID session, and for a dozen people or so, it works well.

When everyone arrives, make these essential announcements: Describe the site and where you will be foraging. Remind them to walk slowly and attentively, meandering on and off the trails into the woods. Invite them to be respectful of the environment and the life at the site.

Tell everyone where and when you will be meeting at the end for the identification session. You might recommend that everyone stay in groups and not wander off alone.

Depending on the experience level of your participants, explain where to find mushrooms (in the dirt and on both living and dead trees) and how to collect a whole specimen, including key identification features that might be on the stipe or just under the ground. A fungus found on wood (or pinecones or animal scat) can be collected with a bit of the substrate it is growing on for identification purposes.

For educational forays, I ask everyone to collect only one or two examples of each unique species, a young one and a mature one if possible, and to not damage trees with their knives or clear the woods of mushrooms. As the sex organs of underground mycelia, mushrooms are more akin to fruit on a tree than to plants. Picking them does not kill the organism, but it's nice to leave some to sporulate later, and for animals and others to enjoy.

If good edible mushrooms are found (or need to be confirmed as such), participants should avoid mixing them with other, possibly poisonous mushrooms. Separate edibles by putting them in paper bags or baskets marked "edible only."

It is safe to touch and handle any mushrooms, even deadly poisonous ones. But you must be sure they are not accidentally ingested, especially by children or animals, and use caution when you do not know the identity of a specimen.

If relevant, point out any toxic plants like poison ivy or giant hogweed in the area and remind foragers to specifically avoid touching them.

Emphasize over and over again to your participants that they should NEVER eat an unidentified specimen and NEVER sample a raw mushroom of any kind in the woods. (See "Is It Edible?" on page 3.)

Display your finds: When everyone returns from foraging, sort out the species on the blanket or the hood of the car. I like to arrange them with similar species and distinguish mushrooms with gills from mushrooms with pores or spines, fungi that look like balls or cups or brackets, and others that don't fit into any of those categories. I also like to include any errant lichen, plants, insect galls, bird's eggs, or bones that end up in participants' baskets. Someone always has something interesting to say about them.

Acknowledge the abundance of life in your woods: See it and marvel at it. Think about ways to care for these environments and to ensure that this abundance can be enjoyed by others in the future.

Interpret your mushrooms: Invite your most knowledgeable participants to talk about the species they know. One person can be assigned to take notes and develop a list for future reference. You can focus on discussing identification features, comparing specimens, describing the culture and names of mushrooms and other items, answering questions, and—with caution—discussing edibility, if known. I find that the most senior mycologists hesitate to discuss edibility with beginners, so as not to give anyone a false sense of skill at recognizing a mushroom, and for reasons of liability. Again, it's better to have a modest and cautious leader and nothing to bring home for dinner than to risk ingesting a poisonous or even deadly mushroom.

Return the mushrooms to the forest: After the finds have been interpreted, and any distributed for further study, drawing, or eating, gather the remaining fungi into a few baskets and spread them around in an unobtrusive (off-trail, less visible) area of the woods. This returns the mushrooms to the forest for other animals and creatures to get nourishment from, and may give the spores a chance to establish themselves in a new area.

Enjoy the company: Bring out the chocolate or a big bag of fresh apples or make hot tea in the woods to make the event something of a seasonal ceremony.

Plums and custard,
*Tricholomopsis
rutilans*

A Mushroomer's Manifesto

1

**"MUSHROOMING" is a word that describes looking for
and collecting mushrooms and other interesting things.**
It belongs, as a legitimate verb, in the common language.
MUSHROOMING can happen in the woods, in the park, in the
supermarket, and in many other surprising environments, like the
basement and the art museum. Sometimes a mushroomer finds
many valuable mushrooms and has plenty of them to cook for dinner.
Sometimes they only find other interesting things.

✦ ✦ ✦

2

MUSHROOMING is a practice of radical stillness.
Collecting a few fungi is not productive and it is not
destructive. The walking is slow, so it can include infirm or
elderly practitioners. It keeps men from racing up mountains.
Small children do it best, when they sit quietly in the dirt just
scanning around. It is what artist and writer Jenny Odell
calls "doing nothing." MUSHROOMING is a place to think,
apart from society, yet completely immersed in the world.

✦ ✦ ✦

3

MUSHROOMING demands uncertainty.
Humility is necessary among mushrooms, especially when you are
concerned with identifying exact species. There are things we will
never know for sure, and what we do know can change as quickly
as the weather. It turns out you can live this way, up to your
neck in mystery, and always open to surprise.

4

MUSHROOMING makes you better at seeing mushrooms and other interesting things.

You become literate in the tiniest details. You can see scales, patches, notches, and the subtlest shifts in coloration, even in the dark. When you see mushrooms, you are looking into a hidden kingdom that is often dismissed and tread upon. You can even see the past in a patch of spore dust and the future in a developing gill.

✦ ✦ ✦

5

MUSHROOMING reminds us to look for beautiful things growing in rot and shit.

In fact, mushrooms remind us that all beautiful things grow in rot and shit, because everything is nourished by the rest of life that came and lived and died before it. "Beauty and grace," Annie Dillard said, "are performed whether or not we will or sense them." You have to show up to see beauty, and you have to look for it.

✦ ✦ ✦

6

MUSHROOMS prove small things contain world-altering powers.

And they should never be underestimated. Something as little as an oysterling or a fairy bonnet or a handful of mycelial thread can evoke a meditation on being, arouse your palate, bring you visions of the divine, or bring hope of renewal to a disturbed environment. It can also kill all the crops and the trees. Or destroy your organs and your mind and precipitate an agonizing death.

7
MUSHROOMS are connected to everything.
Mycelium binds fungi to plants and animals and all life in
the service of nourishment, communication, and decomposition.
We are all alive by a fruitful collaboration of beings—including
mushrooms—across space and time and species. In fact,
everything depends on everything else, and we belong to all of it.

✦ ✦ ✦

8
MUSHROOMS are culture.
There are observations, histories, theories, and new
technologies; there is folklore, cooking, fashion, songs,
art—sometimes even haiku—around a fungus. Like all
of nature, a mushroom is a story we tell about it.

✦ ✦ ✦

9
MUSHROOMS are magic.
"Matter is not lacking in magic, matter is magic,"
Terence McKenna said. And how lucky we are to be alive
to witness it—mushrooms and the rest of the stuff in the
world—in brief and sparkling motion. All of it wondrous.

Recommended Books and Websites

IF YOU WOULD LIKE to develop your mushroom identification skills and learn more about the species near you, look for recent, up-to-date field guides that are specific to mushrooms in your geographic region. Your local mycological association will have some good recommendations for books and websites relevant to your area.

✦ ✦ ✦

Classic technical guides and major mycology websites I would recommend include:

GUIDES

David Arora, *Mushrooms Demystified and All That the Rain Promises and More . . . : A Hip Pocket Guide to Western Mushrooms*

George Barron, *Mushrooms of Ontario and Eastern Canada*

Michael Kuo, 100 *Edible Mushrooms*

Gary H. Lincoff, *National Audubon Society Field Guide to Mushrooms*

Roger Phillips, *Mushrooms and Other Fungi of North America*

Paul Stamets, *Psilocybin Mushrooms of the World: An Identification Guide*

WEBSITES

Learn Your Land (YouTube channel):
youtube.com/c/learnyourland

MushroomExpert.Com (Michael Kuo):
mushroomexpert.com

MykoWeb: Mushrooms, Fungi, Mycology:
mykoweb.com

"Tom Volk's Fungi":
botit.botany.wisc.edu/toms_fungi

Yellow Elanor:
yellowelanor.com

✦ ✦ ✦

If you are interested in learning more about the culture of mushrooming, books I would recommend include:

Eugenia Bone, *Mycophilia: Revelations from the Weird World of Mushrooms*

John Cage, *John Cage: A Mycological Foray*

Peter McCoy, *Radical Mycology: A Treatise on Seeing and Working with Fungi*

Michael Pollan, *How to Change Your Mind: What the New Science of Psychedelics Teaches Us About Consciousness, Dying, Addiction, Depression, and Transcendence*

Merlin Sheldrake, *Entangled Life: How Fungi Make Our Worlds, Change Our Minds and Shape Our Futures*

Paul Stamets (ed.), *Fantastic Fungi: How Mushrooms Can Heal, Shift Consciousness, and Save the Planet*

Anna Lowenhaupt Tsing, *The Mushroom at the End of the World: On the Possibility of Life in Capitalist Ruins*

Long Litt Woon, *The Way Through the Woods: On Mushrooms and Mourning*

✦ ✦ ✦

For more information on some of the artists and artworks discussed in this book

Katie Bethune-Leamen:
katiebethuneleamen.com

Diane Borsato:
dianeborsato.net

David Fenster:
david-fenster.com

Lauren Fournier:
laurenfournier.net

Carsten Höller:
gagosian.com/artists/carsten-holler

Jae Rhim Lee:
studiojaerhimlee.com

Machine Project:
machineproject.com

Kelsey Oseid:
kelzuki.com

Gediminas and Nomeda Urbonas:
nugu.lt/us

Author Acknowledgments

THANK YOU to the many expert mycologists, professional and amateur, who have mentored me in the practice of mushrooming, including Professor Emeritus George Barron, Richard Aaron, the late Gary Lincoff, Paul Sadowski, "Termite" Tim in Guelph, "Mushroom Mike" in Alberta, and Umberto and Linda Pascali. Thank you to the authors of the books on fungi that I have relied on and loved, especially David Arora, Roger Phillips, Michael Kuo, Eugenia Bone, and Merlin Sheldrake. Tremendous gratitude to my Mycological Society of Toronto friends Alan Gan and Vito Testa, and also Duane Sept on the west coast of Canada for reading and doing expert scientific reviews of this work. If there are any leftover errors in my ID notes or descriptions of fungi, they are definitely all mine.

Thank you to the artists featured (and some indirectly) in this book for your work and words, still sporulating in my imagination, including FASTWÜRMS, Katie Bethune-Leamen, Gediminas and Nomeda Urbonas, Jae Rhim Lee, David Fenster, Machine Project, Lauren Fournier, Carsten Höller, Maggie Groat, Dario Ré, JP King for design ideas and moral support, and Andrew Maize for the precious morel spots. And to the late John Cage: I wish we could have been quiet foraging friends.

My deepest appreciation to the illustrator of this volume, Kelsey Oseid—obsessing together every week over paintings of mushrooms, and growing a new friendship with a keen and kindred spirit was unexpected, and it was my favorite part of the process of making this book.

Many thanks to colleagues and friends at the University of Guelph, where my extra-disciplinary dalliances have been not only tolerated but supported with access to an arboretum and an organic farm, expertise, company, and enthusiasm. This includes Chris Earley, Martha Scroggins, Karen Houle, Nathan Saliwonchyk, Madhur Anand, and members of the Guelph Institute for Environmental Research, my coworkers, and my curious, open-hearted students in the department of Studio Art and the College of Arts. Also to others who have invited me to lead forays as a part of their artistic programming, including Ann MacDonald and Erin Peck and the University of Toronto at Scarborough, Jacqueline Bell and Brandy Dahrouge from the Banff Center for the Arts, and Myung-Sun Kim and Clare Butcher of the Toronto Biennial of Art.

Thank you to my friends, especially to Vesna Krstich, Nikolas Drosos, Cynthia Loyst, Jason Tan, Zuzana Eperjesi, Mike Ekers, Leslie Ting, Derek Hamers, Sameer Farooq, Joanne Hui, James Prior, Ruth Comfort, Jamie Girouard, Sarah Lazarovic, Ben Errett, Hazel Meyer, Cait McKinney, Jasmine Oore, Matt Karas, Aida Arnold, Lynda Calvert, Helen Tworkov, Ian and Ruth Sherman, Amanda Trager, Erik Moskowitz, Day Milman, Michael Corrin, and Harriet Alida Lye. Special thanks to my parents, Mary and Mario Borsato, to Dora Collia, and to my in-laws, Anne Morrell Robinson and Joel Robinson, for supporting my preoccupation with fungi (sometimes decaying in baskets all over the porch) over the years with patience, humor, and trust.

Grateful thanks to my literary agent Jackie Kaiser, and also Bridgette Kam from Westwood Creative Artists, for the wise professional advice and for expertly guiding this book into the world. Many thanks to the staff at Douglas & McIntyre, including Diane Robertson for her outstanding graphic design, and especially to editor Artie Goshulak and copy editor Merrie-Ellen Wilcox for kind words and your caring, meticulous work. Heartfelt thanks to publisher Anna Comfort O'Keeffe for the pine mushroom spots and the exceptionally generous support of this book and of my writing.

Thank you to the Canada Council for the Arts, whose financial support of this project—and of my art practice over the years—has made my work possible.

Finally, to my partner, Amish Morrell, who showed me my first chanterelle and my son, Felix, who comes along, even in the rain: thank you. You are my favorite fairy people.

—Diane Borsato

Illustrator Acknowledgments

THANK YOU to my husband, Nick, and young son, Aiden, who accompanied me into the woods several times and laughed, clapped, and jumped up and down with me when I found my first pinwheel mushroom. And, oh, to be rapturously art-directed by a passionate mind like Diane Borsato's! Her mushrooming verve is palpable in every interaction, and spread to me like wildfire as we fastidiously, joyfully re-created these species in gouache over the course of several months.

—Kelsey Oseid

Index

Strobilomyces strobilaceus; see old man of the woods (*Strobilomyces floccopus*)
Stropharia aeruginosa (verdigris agaric), 188–89, **188**
Stropharia rugosoannulata (wine cap), 188–89, **189**
Urbonas Studio, 208–10
Suillus luteus (slippery jack), 140, **142**
summer truffle (*Tuber aestivum*), 20
supermarket mushrooms, 80

T

Taleggio, 23
Testa, Vito, 14, 102
tinder conk; *see* hoof fungus (*Fomes fomentarius*)
tippler's bane (*Coprinopsis atramentaria*), 179, **179**
Trametes versicolor (turkey tail), 161, **163**
Tranströmer, Tomas, 56
Tricholoma magnivelare (matsutake), 52–53, 130–33, **131**
Tricholoma murrillianum (matsutake), 52–53, 130–33, **131**
Tricholomopsis rutilans (plums and custard), 166, **167, 215**
trompettes de la mort ("trumpets of death"), 24, **25**
truffle (*Tuber* spp.), 20–23, 53
"trumpets of death" (*trompettes de la mort*), 24, **25**
Tuber aestivum (summer truffle), 20
Tuber magnatum (white truffle), 20, 53
Tuber melanosporum (black truffle), 20–23, **21**, 53
turkey tail (*Trametes versicolor*), 161, **163**
Tylopilus chromapes; see sleeping beauty (*Harrya chromapes*)
Tylopilus felleus (bitter bolete), 108, **109**

U

Urbonas, Gediminas, 12, 208–10
Urbonas, Nomeda, 12, 208–10
Urnula craterium (devil's urn), 170, **173**

V

Valentová, Tereza, 79
Vandeleur-Boorer, Alice, 79
velvet foot; *see* enoki (*Flammulina velutipes*)
verdigris agaric (*Stropharia aeruginosa*), 188–89, **188**
Vermilion Sands, 209
Verpa bohemica (early morel), 202

Verpa conica (half-free morel), 202
violet coral (*Clavaria zollingeri*), 30, **33**
Volk, Tom, 182
Volvariella volvacea (straw mushrooms), 53

W

Wallace, David Foster, 168
Warhol, Andy, 136
Wasson, Gordon, 114
Wasson, Valentina, 114
white baneberry; *see* doll eyes (*Actaea pachypoda*)
white elfin saddle (*Helvella crispa*), 190, **191**
white truffle (*Tuber magnatum*), 20, 53
wine cap (*Stropharia rugosoannulata*), 188–89, **189**
winter truffle; *see* black truffle (*Tuber melanosporum*)
witch's butter (*Dacrymyces chrysospermus*), 192, **193**
witch's hat (*Hygrocybe conica*), 194–96, **195**
wolf's milk slime mold (*Lycogala epidendrum*), 69, **69**
wood ear (*Auricularia americana*), 52, 197–99, **198**
Woon, Long Litt, 117

X

Xylaria polymorpha (dead man's fingers), 56, **57**

Y

yellow common jelly baby (*Leotia lubrica*), 97
yellow houseplant mushroom; *see* plant pot dapperling (*Leucocoprinus birnbaumii*)
yellow morel (*Morchella esculenta*), 200–2, **200**
yellow stainer (*Agaricus xanthodermus*), 80, **83**
yellow swamp brittlegill (*Russula claroflava*), 204, **205**
yellow-foot chanterelle (*Craterellus tubaeformis*), 45, **46**

Z

"Zooetics," 208–10
Zooetics Pavilion of Ballardian Technologies, **208**, 210

About the Author

Diane Borsato is an award-winning artist, naturalist, and educator. In her work she has explored our relationships to nature in collaboration with a range of practitioners, including beekeepers, mushroomers, astronomers, falconers, and orchardists. She has exhibited at the Art Gallery of Ontario, the Museum of Contemporary Art, the National Art Center, the Toronto Biennial of Art, the Vancouver Art Gallery—and in many other galleries and museums around the world. She was the coeditor and a contributor to the book *Outdoor School Contemporary Environmental Art* (D&M, 2021) with Amish Morrell, and is an associate professor of studio art at the University of Guelph. She lives in Toronto, Ontario.

dianeborsato.net | ⓘ dianeborsato

About the Illustrator

Kelsey Oseid is an illustrator, author, and amateur naturalist. Her gouache illustrations focus on natural history subjects like taxonomy and biodiversity, as well as related subjects like astronomy and the ways humans relate to the natural world. She is the author and illustrator of three books with Ten Speed Press. She lives with her husband and son in Minneapolis, Minnesota.

kelzuki.com | ⓘⓨ kelzuki